# RIP-ET: A Riaparian Evapotranspiration Package for MODFLOW-2005

By Thomas Maddock III and Kathryn J. Baird, University of Arizona; R.T. Hanson, U.S. Geological Survey; and Wolfgang Schmid and Hoori Ajami, University of Arizona

Chapter 39 of
**Section A, Groundwater**
**Book 6, Modeling Techniques**

Office of Groundwater, Transboundary Aquifer Assessment Program

Techniques and Methods 6–A39

U.S. Department of the Interior
U.S. Geological Survey

**U.S. Department of the Interior**
KEN SALAZAR, Secretary

**U.S. Geological Survey**
Marcia K. McNutt, Director

U.S. Geological Survey, Reston, Virginia: 2012

For more information on the USGS—the Federal source for science about the Earth, its natural and living resources, natural hazards, and the environment, visit http://www.usgs.gov or call 1–888–ASK–USGS.

For an overview of USGS information products, including maps, imagery, and publications, visit http://www.usgs.gov/pubprod

To order this and other USGS information products, visit http://store.usgs.gov

Suggested citation:
Maddock, Thomas, III, Baird, K.J., Hanson, R.T., Schmid, Wolfgang, and Ajami, Hoori, 2012, RIP-ET: A riparian evapotranspiration package for MODFLOW-2005: U.S. Geological Survey Techniques and Methods 6-A39, 76 p.

# Contents

# Figures

# Tables

# Conversion Factors

## Inch/Pound to SI

| Multiply | By | To obtain |
|---|---|---|
| foot (ft) | 0.3048 | meter (m) |
| foot per day (ft/d) | 0.3048 | meter per day (m/d) |
| foot per second (ft/s) | 0.3048 | meter per second (m/s) |
| square foot ($ft^2$) | 0.09290 | square meter ($m^2$) |
| square mile ($mi^2$) | 2.590 | square kilometer ($km^2$) |

## SI to Inch/Pound

| Multiply | By | To obtain |
|---|---|---|
| centimeter (cm) | 0.3937 | inch (in.) |
| centimeter per day (cm/d) | 0.03281 | foot per day (ft/d) |
| kilometer (km) | 0.6214 | mile (mi) |

# Acknowledgments

This research was supported in part by two grants. The first was an Environmental Protection Agency-National Science Foundation (EPA-NSF) Star Grant entitled "Restoring and Maintaining Riparian Integrity in Arid Watersheds," number R827150/01, which funded field experiments in the South Fork of the Kern River in California and the San Pedro River in Arizona. The second was a National Center for Environmental Assessment (NCEA) Global Change Division CO-OP Agreement entitled "Potential Effects of Climate Change on Biodiversity and Riparian Wildlife Habitat of the Upper San Pedro River," which funded development of this software package. The views and conclusions contained in this document are those of the authors and should not necessarily be interpreted as representing official policies (expressed or implied) of the EPA or NCEA.

RIP-ET has evolved from experiences in application to MODFLOW models of the South Fork Kern River Valley, California; the Lower Rio Grande Basin, New Mexico; Arivaca Watershed, Arizona; the Upper San Pedro River Basin, Arizona; and the Colorado River Delta, Mexico. We also have had written and verbal communications from various users throughout the United States and abroad. Coding help has come from Arlen Harbaugh and Stan Leake of the U.S. Geological Survey.

# RIP-ET: A Riparian Evapotranspiration Package for MODFLOW-2005

By Thomas Maddock III[1], Kathryn J. Baird[1], R.T. Hanson[2], Wolfgang Schmid[1], and Hoori Ajami[1]

## Abstract

A new evapotranspiration package for the U.S. Geological Survey's groundwater-flow model, MODFLOW, is documented. The Riparian Evapotranspiration Package (RIP-ET) provides flexibility in simulating riparian and wetland transpiration not provided by the Evapotranspiration (EVT) or Segmented Function Evapotranspiration (ETS1) Packages for MODFLOW 2005. This report describes how the RIP-ET package was conceptualized and provides input instructions, listings and explanations of the source code, and an example.

Traditional approaches to modeling evapotranspiration (ET) processes assume a piecewise linear relationship between ET flux and hydraulic head. The RIP-ET replaces this traditional relationship with a segmented, nonlinear dimensionless curve that reflects the eco-physiology of riparian and wetland ecosystems. Evapotranspiration losses from these ecosystems are dependent not only on hydraulic head, but on the plant types present. User-defined plant functional groups (PFGs) are used to elucidate the interaction between plant transpiration and groundwater conditions. Five generalized plant functional groups based on transpiration rates, plant rooting depth, and water tolerance ranges are presented: obligate wetland, shallow-rooted riparian, deep-rooted riparian, transitional riparian and bare ground/open water. Plant functional groups can be further divided into subgroups (PFSGs) based on plant size, density or other characteristics.

The RIP-ET allows for partial habitat coverage and mixtures of plant functional subgroups to be present in a single model cell. RIP-ET also distinguishes between plant transpiration and bare-ground evaporation. Habitat areas are designated by polygons; each polygon can contain a mixture of PFSGs and bare ground, and is assigned a surface elevation. This process requires a determination of fractional coverage for each of the plant functional subgroups present in a polygon to account for the mixture of coverage types and resulting transpiration. The fractional cover within a cell has two components: (1) the polygonal fraction of active habitat (excluding area of bare ground, dead trees, or brush) in a cell, and (2) fraction of plant type area or bare ground area in a polygon. RIP-ET determines the transpiration rate for each plant functional group and evaporation from bare ground/open water in a cell, the total ET in the cell, and the total ET rate over the region of simulation.

## 1.0 Introduction

The physical settings and dynamic processes of streams have changed dramatically over the past century (Stromberg, 2001). For example, in the Southwestern United States, many of the characteristic desert riparian habitats of cottonwood-willow galleries, ciénegas (warm temperature wetlands) and mesquite bosques (wooded area) that once lined the rivers are degrading and disappearing or have disappeared (Glennon and Maddock, 1994). The dynamics of riparian and wetland ecosystems are closely tied to groundwater and streamflow hydrology (Busch and others, 1992; Grimm and others, 1997). The pattern and quantity of surface and subsurface flows are the primary determinants of riparian ecosystem structure and function, although depth to groundwater and its rate of change place important constraints on the distribution and vigor of riparian vegetation types (Stromberg and others, 1996; Poff and others, 1997; Shafroth and others, 2000). Water commonly is the principal limiting resource for these critical systems; surface-water diversions and groundwater extractions often contribute to habitat degradation and loss (Stromberg, 1993).

---

[1] Department of Hydrology and Water Resources, University of Arizona.

[2] U.S. Geological Survey.

Riparian and wetland systems contain a disproportionate share of regional biodiversity and play an important role in the regional water and energy balance, especially in arid and semi-arid regions (Williams and others, 1998). If these ecologically significant systems are to be preserved, the remaining systems and water resources must be carefully managed. Reliable and accurate methods for estimating the requirements for preserving and restoring these systems are essential. Regional groundwater models are a promising tool for this estimation and analysis. However, if regional groundwater models are to be used, accurate estimates of boundary conditions are required. One of the most critical but poorly quantified groundwater boundary conditions is seasonal riparian evapotranspiration (Goodrich and others, 2000). When modeling the flow in these systems, the method by which evapotranspiration (ET) is simulated can affect calculated heads and thus the interpretations regarding system dynamics (Banta, 2000).

In many groundwater models, stream and evapotranspiration are simulated as temporally varying inflows and outflows. Traditionally, the ET sink term is treated as head-dependent in a piecewise linear fashion that monotonically approaches a maximum ET rate (Banta, 2000; Harbaugh, 2005). Although this quasi-linear relationship may hold true for evaporation, it does not accurately reflect the relationship between riparian transpiration and groundwater conditions. RIP-ET was developed to simulate evapotranspiration from the water table that lies beneath riparian/wetland systems in a manner that reflects their ecology and physiology (Maddock and Baird, 2003; Schorr, 2005; Barth and others, 2008). RIP-ET can be used with other MODFLOW processes in adjacent regions, such as the Farm Process (FMP) (Schmid and others, 2006; Schmid and Hanson, 2009) for simulating ET from agriculture and native vegetation. The RIP-ET source codes for MODFLOW-2005 are presented in this report.

# 2.0 Plant Functional Groups

Riparian and wetland ecosystems are composed of various plant types and species. The identification and use of plant functional groups can assist in reducing the enormous complexity of individual species and populations into a relatively small number of general recurrent patterns. This technique has emerged as a useful way to organize plant species that have similar impacts on ecosystem processes into manageable and meaningful categories (Leishman and Westoby, 1992; Williams and others, 1998). Plant functional groups are defined as non-phyllogentic groupings of plant species that exhibit similar responses to environmental conditions and have similar effects on the dominant ecosystem processes (Lavorel and others, 1997). In RIP-ET, plant functional groups are used to elucidate the interactive processes of plants (and plant ET) with groundwater conditions. The model is designed to be flexible; the user determines which sets of plant functional groups are appropriate for the simulation and geographic region of the riparian/wetland system to be modeled.

In the development of the RIP-ET, we defined five basic functional groups based on transpiration rates and processes, plant rooting depth, and drought tolerance. Although the groups presented here are for semi-arid environments, the methodology can be applied universally. The generalized plant functional groups are: obligate wetland, shallow-rooted riparian, deep-rooted riparian, transitional riparian, and bare ground/open water. Bare ground/open water, although not a plant functional group, must be included to accurately simulate evapotranspiration from the cell or active model area.

The obligate wetland plant functional group contains plants that require saturated soil conditions or standing water. Throughout most of the United States, species in this category are herbaceous and generally have shallow root systems. For example, in the Southwestern United States, species such as *Typha* spp. (cattail), *Scirpus* spp. (bulrush), and *Juncus* spp. (rushes) typify this group.

Riparian forests are composed of drought-intolerant phreatophytes that rely on shallow groundwater for establishment, growth and transpiration (Busch and others, 1992; Stromberg, 1993; Snyder and Williams, 2000). Due to the difference in rooting depths and transpiration rates, shallow-rooted and largely herbaceous riparian species [such as *Xanthium* sp. (cocklebur) and *Rumex crispus* (curly dock)] are categorized separately from deep-rooted riparian plants such as *Populus* spp. (cottonwoods) and *Salix* spp. (willows). In some cases, multiple deep-rooted categories may be useful, depending on the species present. For example, this refinement allows the distinction of *Prosopis velutina* (bosque mesquite trees) from cottonwood and willows. Unlike obligate wetland species, neither the shallow nor the deep-rooted riparian group tolerates saturated conditions for extended periods of time.

The last plant functional group, transitional (or facultative) riparian, consists of species that although not strictly dependent on a high water table, have water requirements that generally exceed the surrounding environment. These species generally live along the outer edges of riparian systems and include *Sporobolus Wrightii* (sacaton), *Sambucus* sp. (elderberry), *Juglans* spp. (walnut), *Celtis* sp. (hackberry), and *Platanus* sp. (sycamore). The transpiration rates and range of depth-to-groundwater over which these plant functional groups thrive differ for each group (fig. 1), as will be discussed below.

# 3.0  Plant Functional Subgroups

Plant size and density play roles in determining transpiration rates. Large woody plants have different maximum rooting depths, hydraulic architecture and transpiration rates than smaller trees (Meinzer and others, 1997). Furthermore, areas with dense configurations of woody and herbaceous plants may show increased transpiration rates compared to sparse configurations. Taken together, the plant functional group and plant size or density defines a plant functional subgroup (PFSG). Examples of possible plant functional subgroups are: transitional riparian-large or wetlands-medium density. Table 1 presents possible plant functional subgroups for this example. This table is by no means exhaustive and is meant to be illustrative.

Transpiration rates and the range of depth to groundwater over which these groups live differ for each plant functional group (fig. 1). Streamside riparian corridors typically contain multiple plant functional subgroups, each reacting differently to the hydrologic conditions. A deeper water table may be ideal for a transitional riparian group, but near fatal to a wetland or herbaceous, shallow-rooted group. Similarly, a high water table conducive to a obligate wetland group may drown out the transitional riparian group. Even so, a MODFLOW cell covering a riparian system likely will include a number of compatible plant functional groups. Thus, the transpiration from a riparian/wetland system is dependent on the plant functional subgroups present and on the groundwater depth.

**Table 1.**  Illustrative subgroupings for plant functional groups, by size and density.

| Plant group | Size | Densities | Species examples |
|---|---|---|---|
| Obligate wetland | N/A | Low, medium, high | Cattail, Bulrush |
| Shallow-rooted riparian | N/A | Low, medium, high | Curly dock, Cocklebur, Deer grass |
| Deep-rooted riparian | Small, medium, large | Low, medium, high | Cottonwood, Willow, Mesquite (bosque) |
| Transitional riparian | Small, medium, large | Low, medium, high | Walnut, Hackberry, Sacaton, Sycamore |

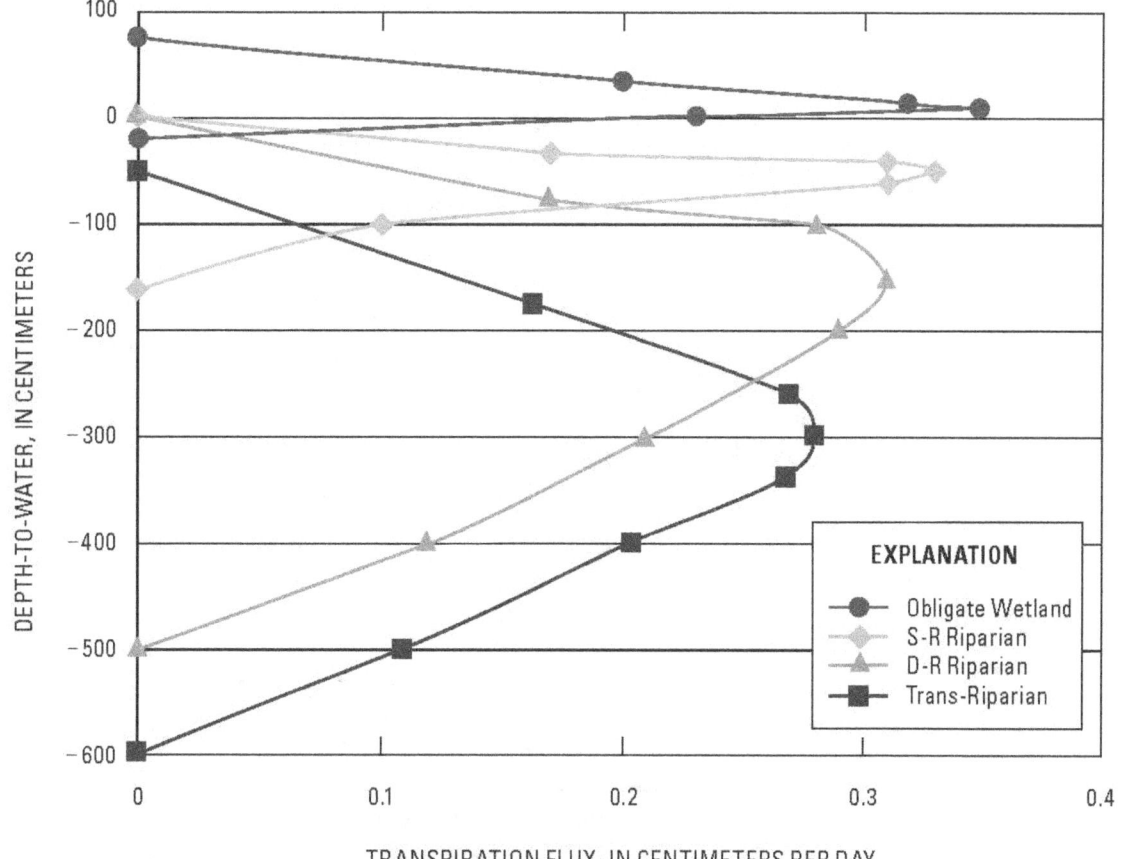

**Figure 1.**  Average transpiration flux curves for four plant functional groups—obligate wetland, shallow rooted (S-R) riparian, deep-rooted (D-R) riparian and transitional (Trans) riparian (modified from Baird and Maddock, 2005).

# 4.0 Transpiration Flux Curves

The transpiration rate per area is called the transpiration flux. Due to differences in sensitivity to water availability and rooting depth, transpiration fluxes in plant species, and subgroups, vary with groundwater depths. Based on a literature review, field measurements, and research, Baird and Maddock (2005) developed preliminary transpiration flux curves for the basic set of plant functional groups (fig. 1). These curves provide the transpiration flux as a function of depth to the water table.

Figure 2 illustrates a representative transpiration flux curve. For each plant functional subgroup, there is a water-table elevation or equivalent elevation of the extinction depth (Hxd in fig. 2) below which the roots can not obtain water and ET is nonexistent. When the water table rises and water becomes available to the root system, the transpiration rates increase until they reach an average maximum ET flux (Rmax). At high water-table elevations, the root systems become oxygen deficient and transpiration rates decrease until the plants die of anoxia. The water-table elevation associated with plant death is the saturated extinction depth elevation (Hsxd in fig. 2). Although the decrease in transpiration due to anoxia or transpiration from flooded landscapes is included in FMP (Schmid and others, 2006; Schmid and Hanson, 2009), the decrease in transpiration flux resulting from a high water table has not been considered in previous MODFLOW evapotranspiration packages (McDonald and Harbaugh, 1988; Harbaugh and McDonald, 1996a; Banta, 2000). Although figure 2 shows Hsxd below the land-surface elevation, it need not be; it can be at or above the land surface for some subgroups. For example, some wetland species may tolerate as much as a meter of standing water above the land surface.

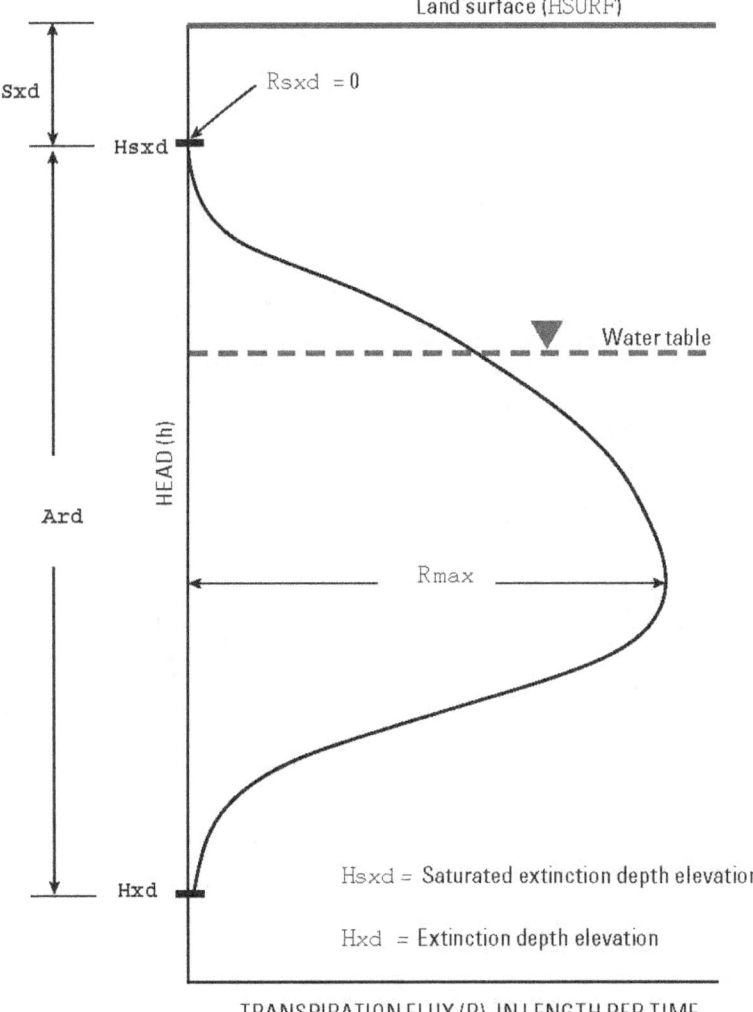

**Figure 2.** Generic transpiration flux curve (modified from Baird and Maddock, 2005).

The distance between the two extinction depth elevations is called the active root zone depth (Ard). The saturated extinction depth (Sxd) is measured with respect to the land-surface elevation, HSURF. *If* Hsxd is below the land-surface elevation, Sxd is positive; if it is above, Sxd is negative. The following relations hold:

$$Hsxd = HSURF - Sxd$$

and

$$Hxd = Hsxd - Ard.$$

The user of the RIP-ET package inputs the name (RIPNM) and the values of Sxd and Ard for each plant functional subgroup. The user also inputs the maximum transpiration flux (Rmax) and the transpiration flux at Hsxd (Rsxd) for each plant functional subgroup (see below). These values must be in units consistent with the MODFLOW simulation. The maximum transpiration flux (Rmax) is the average head-dependent transpiration rate, not a peak transpiration rate. The Rmax should be averaged over the entire plant-group polygon for the periods of time that are consistent with the model stress periods. For plant transpiration, Rsxd = 0. For bare ground or open water, Rsxd equals the evaporation rate, and a shallow, more appropriate, extinction depth should be applied. If the water table is at or very near the land surface, evaporation will be non-zero even if plants are dead and transpiration is zero.

# 5.0  Transpiration Flux Curve Linear Interpolation

RIP-ET does not use the continuous curve of figure 2, but instead uses an approximation based on linear segments as illustrated in figure 3. The transpiration fluxes reported in the literature are in various units [for example, cm/s, ft/d, (L/m)/d, (kg/m$^2$)/s]; the latter units need to be adjusted by the density of water to be dimensionally correct). To help alleviate problems that might occur with a particular choice of units, the curve segments are read into the RIP-ET as dimensionless. They are internally converted to units consistent with other MODFLOW packages based on the values of Rmax and Ard. The Rmax (L/T) and Ard (L) units must match the length and time units used elsewhere in the simulation.

The process for producing the dimensionless segmented curves is conducted externally to the RIP-ET package by the user, and proceeds as follows. Figures 4 and 5 present the hypothetic, segmented transpiration flux curve. Segments are defined by vertices ($h(k)$, $R(k)$) and determine the shape of the curve. In figure 4, $d(1)$, $d(2)$...$d(N)$, are in length units and represent the change in head over a segment, although in figure 5, $dR(1)$, $dR(2)$... $dR(N)$ are in flux units and represent the change in flux over a segment. The indices and segments must start at Hxd, the extinction depth elevation. If $h(1)$ and

$R(1)$ are the head and transpiration flux at the extinction depth elevation; and $h(2)$ and $R(2)$ are the head and transpiration flux at the end of the first segment, then the $d$'s and $dR$'s are defined as,

$$d(1) = h(2) - h(1)$$

and

$$dR(1) = R(2) - R(1).$$

In general, for $N$ segments and $1 \le k \le N$, where the numbering of $k$ moves upward toward the saturated extinction depth,

$$d(k) = h(k+1) - h(k)$$

and

$$dR(k) = R(k+1) - R(k).$$

For a functional subgroup, one must define the following dimensionless segment variables that define the proportion of the active root zone depth represented by $k^{th}$ segment,

$$fdh(k) = \frac{d(k)}{Ard} \qquad (5.1)$$

and

$$fdR(k) = \frac{dR(k)}{Rmax}. \qquad (5.2)$$

Note that for the $N$ segments,

$$\sum_{k=1}^{N} fdh(k) = 1$$

and

$$\sum_{k=1}^{N} fdR(k) = \frac{Rsxd}{Rmax}$$

In figure 4, Rsxd = 0, but, as previously mentioned, Rsxd can be non-zero for evaporation curves such as those for open water or bare land. In these cases, Rsxd is equal to Rmax and sum of the fdR($k$) over all the segments is one.

The number of segments (NuSeg) for each plant functional subgroup, and the dimensionless segment variables (fdh($k$) and fdR($k$)) are input for each segment and each plant functional subgroup by the RIP-ET package user. Example values for NuSeg, fdh($k$) and fdR($k$) for four of the plant functional groups and three deep-rooted riparian size classes for the South Fork of the Kern River, California, are shown in appendix A.

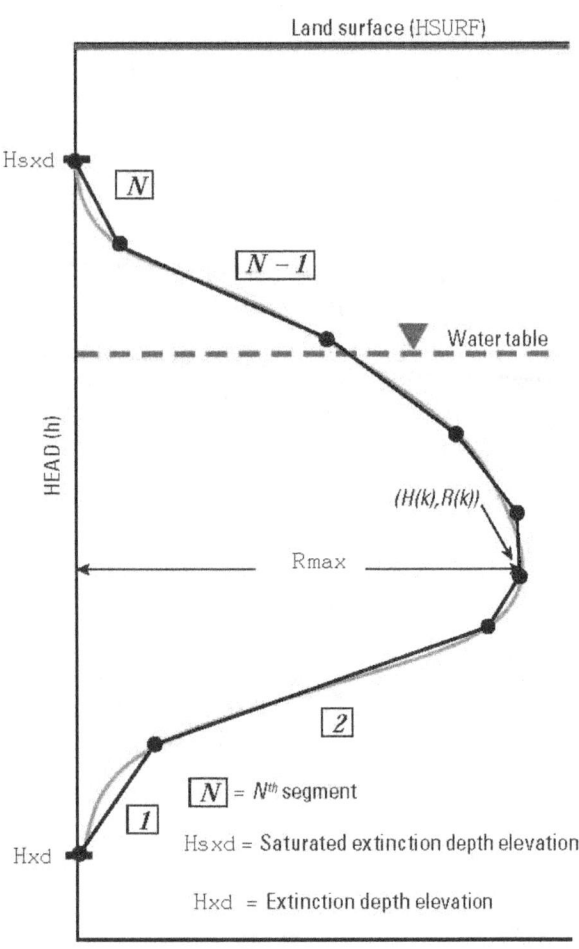

**Figure 3.**  Hypothetical segmented transpiration flux curve.

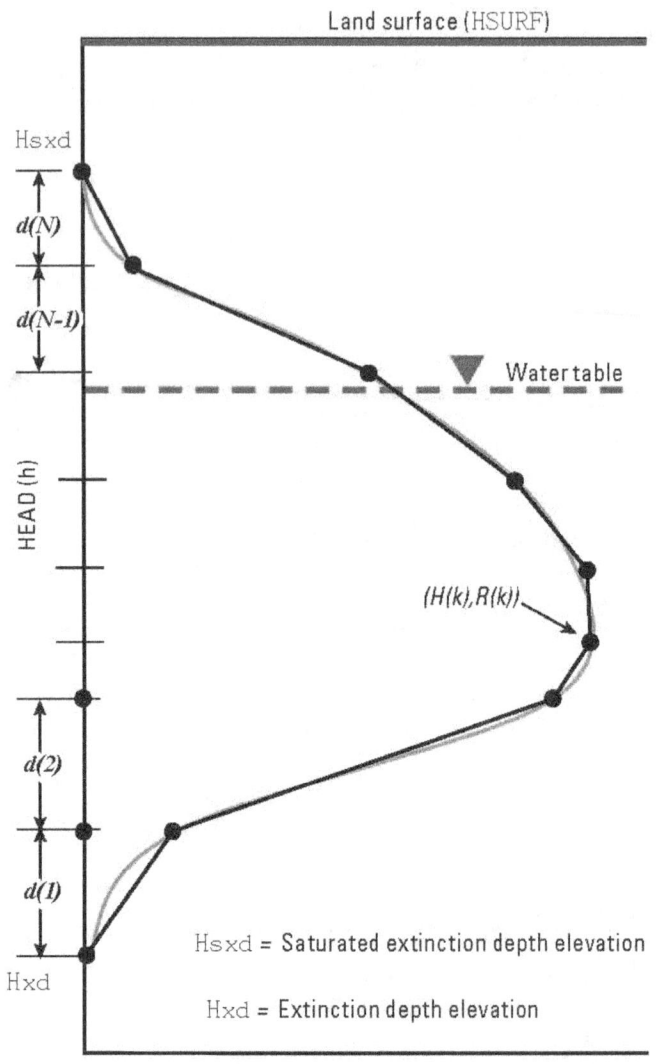

**Figure 4.**  Linear interpolation of transpiration flux—*d's*.

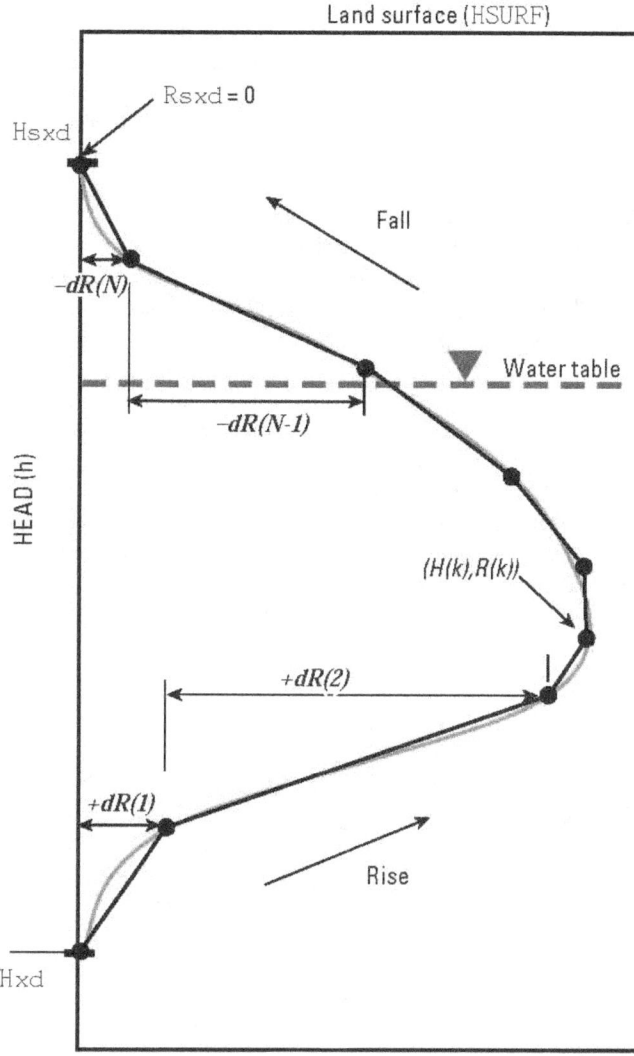

**Figure 5.** Linear interpolation of transpiration flux—*dR*'s.

# 6.0 Development of Plant Functional Group Curves

There are two steps involved in developing a transpiration flux curve: (1) determining the extinction and saturated extinction depths, and (2) determining the shape of the transpiration curve. Extinction depths can be approximated from the maximum rooting depth(s) of the species or plant group in question. These values can be obtained through field studies, literature research, or a combination of both. Saturated extinction depths are a bit more problematic, as flood depth tolerances have been measured in only a small number of species. However, upper limits can be extrapolated from correlative studies between plant species occurrence and

water-table elevations (Vreugdenhil and others, 2006). These saturated extinction depths may not be precise but provide a good starting point by representing the upper and lower limits of water tolerance for the species or plant functional subgroups.

The shape of a transpiration curve is dependent on the relationship between transpiration rate and water-level depths. To determine this, one needs estimated or measured transpiration values at different water depths. Currently, there are three broad methods or categories used to calculate transpiration fluxes. The first category of methods involves measuring transpiration using either stomatal conductance measurements or measured velocities of sap flow in the vascular tissue in individual trees (Williams and others, 1998; Steinwand and others, 2001). These measurements must then be scaled up to the stand size. A common approach to scale flux measurements of woody plant transpiration (deep-rooted riparian and transitional riparian trees) to stand-level water use, relies on obtaining mean flux and a mean canopy-to-sapwood area ratio at the population level (Oren and others, 1999; Granier and others, 2000). A canopy transpiration flux is defined as,

$$\text{Canopy transpiration flux rate} = \frac{\text{Xylem sap velocity} \times \text{Sapwood area}}{\text{Canopy area}}.$$

A similar approach within this category relies on the relationship of sapwood to tree diameter at breast height (DBH) or tree basal area. In this instance, the head dependent flux is multiplied by the fraction of basal area per subgroup area. If ET fluxes are measured as a flux per tree or tree within a size class, then the scaling factor should be density (that is, number of trees) per unit canopy area.

The second category of methods involves measuring integrated evaporation and transpiration fluxes from micrometeorological measurements such as Eddy-covariance and Bowen-ratio and then trying to parse the integrated flux into species or plant functional group rates. Eddy-covariance flux towers commonly are used on rivers in the Western United States. Numerous studies have quantified transpiration for riparian areas and phreatophytic scrub communities using the technique (Goodrich and others, 2000; Berger and others, 2001; Scott and others, 2004). Eddy-covariance and Bowen-ratio flux towers appear to produce the best estimates of evapotranspiration rates at scales of 0.1–1 km (Rana and Katerji, 2000).

The third category of methods involves an increasing number of remote sensing techniques. To determine fluxes at the landscape or regional scale, various remote sensing techniques or combined eddy covariance with remote sensing are used (Nagler and others, 2005; Kimura and others, 2007). These methods take advantage of such tools as MODIS (Moderate Resolution Imaging Spectometer) on the NASA Terra satellite.

The methodology used to measure transpiration generally depends on the type of plant functional subgroup of interest, the corresponding terrain, the scale and discretization of the model, and the researcher's objective. Depending on the specific distribution of plant groups, the approach to creating relations of ET to plant functional groups (PFGs) needed to simulate transpiration may require additional measurements of the specific plant groups to supplement the integrated ET canopy measurements.

# 7.0 Plant Functional Subgroup Coverage

The transpiration flux curves discussed in sections 6.0 and 7.0 provide the discharge/area for the transpiration processes in a cell. To determine the discharge for each plant functional subgroup (PFSG), the transpiration flux is multiplied by the area covered by that plant functional subgroup. This area is referred to as the plant functional subgroup coverage area, or more simply, as PFSG coverage.

The use of this package requires having information about the distribution of the plant functional subgroups within the active riparian areas in the model domain. Depending on cell size, each model cell is unlikely to contain 100 percent active riparian or wetland habitat; nor is a cell likely to contain all possible plant functional subgroups. Furthermore, the PFSG coverages may change from cell to cell. Therefore, RIP-ET requires a fractional coverage value for each of the plant functional subgroups present in a cell to simulate the mixture. Each fractional coverage value is then multiplied by the surface area of the cell to provide the cell discharge area for that plant functional subgroup.

Plant coverage information is organized by polygons. A polygon contains a single or mixture of plant functional subgroups and has an approximately uniform land-surface elevation throughout. Aerial photographs, digital elevation models (DEMs), and field studies can be used to delineate the polygon and associated properties. Polygon formulation and plant coverage determination is an operational procedure and is somewhat subjective. Assistance from a plant ecologist about local conditions may prove highly beneficial.

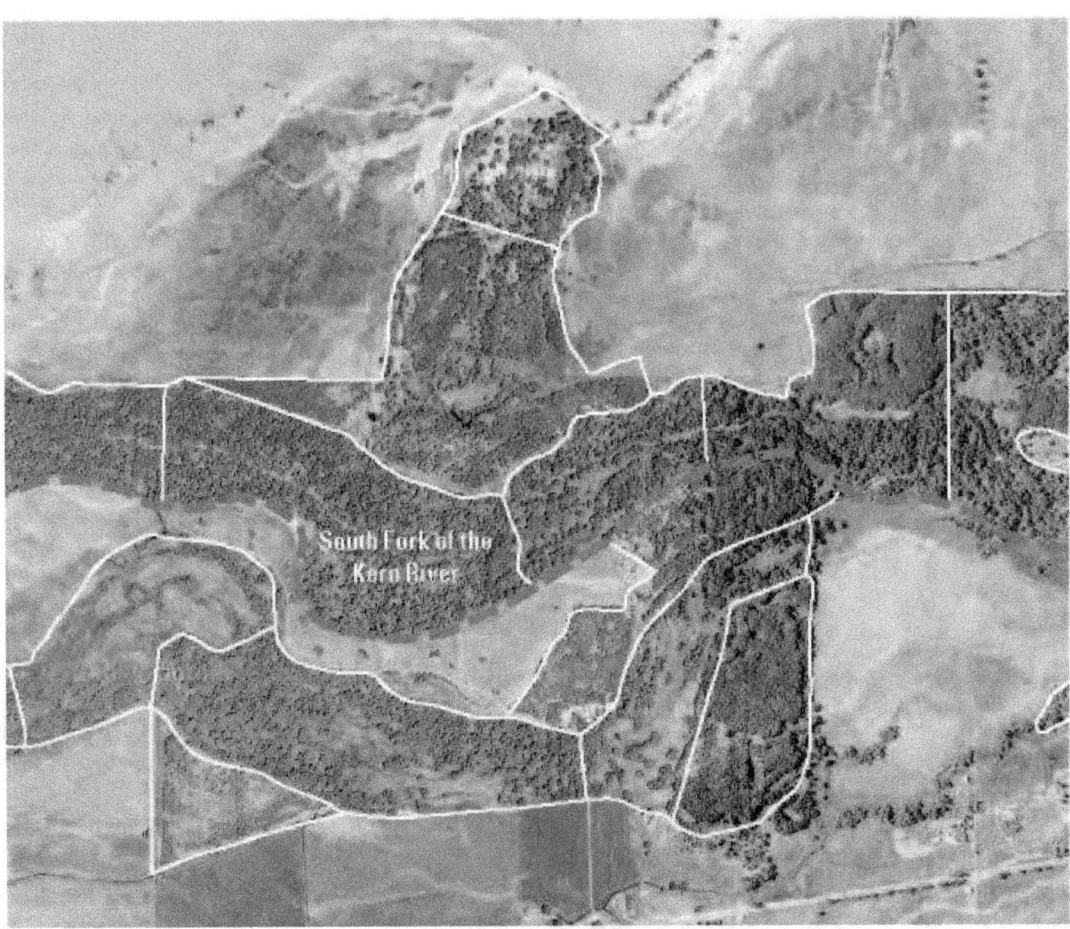

**Figure 6.**   Riparian polygons defined for a portion of the South Fork of the Kern River, California. (Dashed line is Kern River and white outlined polygons represent habitat subregions of plant subgroups.)

Bare ground areas not underlain by a shallow water table (for example, the non-vegetated area in fig. 6) can be excluded from the cell area by not being incorporated into a polygon. Bare ground areas underlain by a shallow water table must be included within polygons to allow for evaporation.

One of the benefits of RIP-ET is its ability to deal with large quantities of detailed physical data—plant coverages and land-surface elevations. To aid the user in preparing the datasets for RIP-ET, a GIS (Geographic Information System) based preprocessor has been developed to guide the user through the process of preparing the required data files (Ajami and others, 2011).

Figure 6 shows a set of polygons defined for a portion of the riparian area along the South Fork of the Kern River in California. Thereafter, a MODFLOW grid is superimposed over the mosaic of photographs that constitute the visual representation of the riparian area (fig. 7). The grid and the mosaic must be properly aligned and registered. The superposition of the grid defines a new set of polygons that may have MODFLOW cell boundaries delimiting them. As shown in figure 7, each cell with riparian habitat contains one or more of the new polygons. The hachured region is an example of one of the new polygons.

For each cell that contains riparian habitat, the number of polygons is recorded in the variable NPOLY(i), indexed by a cell number. For example, for the $i^{th}$ cell in figure 7, there are seven polygons.

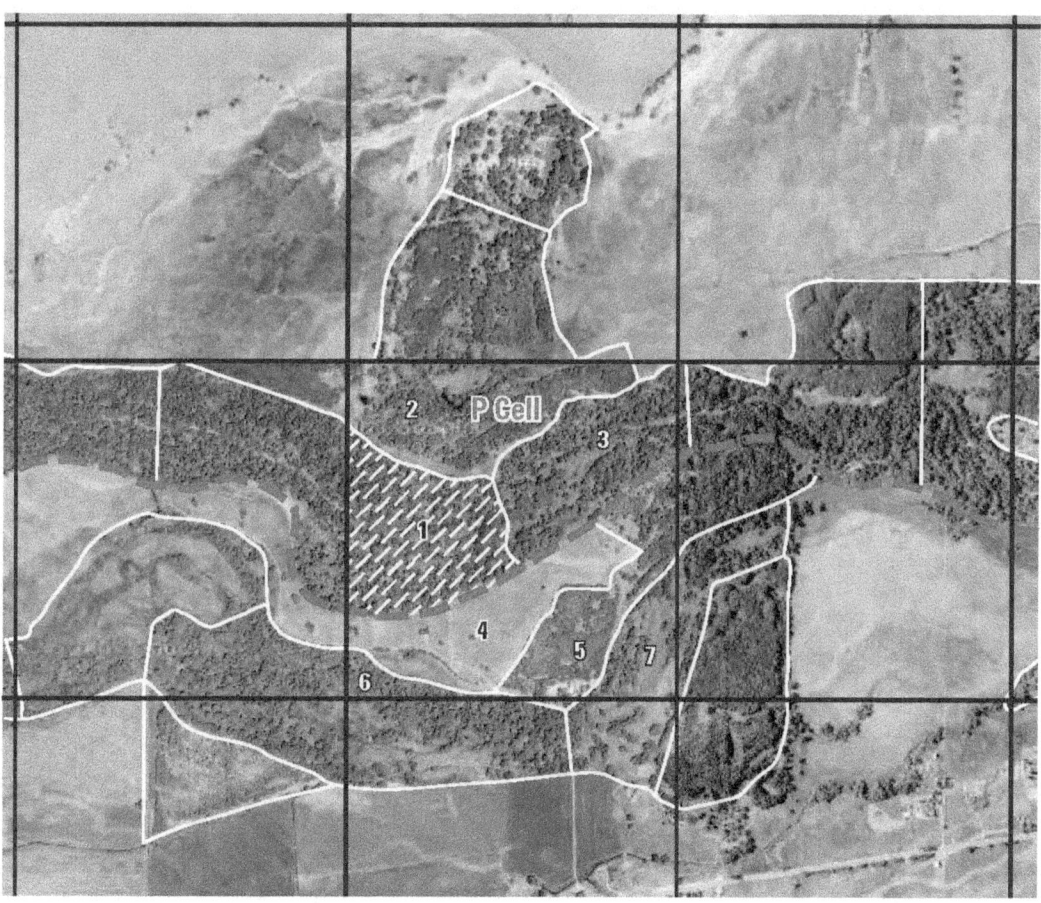

**Figure 7.**  A portion of the MODFLOW grid (black squares) superimposed over the photo-mosaic of the South Fork of the Kern River, California, that creates new sets of plant-group polygons (white outlined polygons), one of which is hachured and labeled P cell. (Dashed line is Kern River.)

## 7.1  Polygonal Fraction of Active Habitat in a Cell

The polygonal fraction of active habitat in a cell refers to the ratio of the area of a polygon in a cell to the total area of the cell. Because cell sizes in MODFLOW may be quite large (for example, square kilometer), **individual cells may contain all or parts of numerous polygons containing riparian habitat** (see fig. 7). The fraction of active habitat in polygon $j$ for cell $i$ ($fPA(i,j)$) is defined as,

$$fPA(i, j) = \frac{\text{area of polygon } j \text{ for cell } i}{\text{area of cell } i}.$$

## 7.2  Fraction of Plant Flux Area in a Polygon

Within any riparian habitat polygon, the area associated with each of the transpiration flux curves rarely equals 100 percent, although for multi-layered environments, the total percentage of plant or canopy cover within a polygon can exceed 100 percent. At issue is: what are the actual flux areas that are contributing to each type of transpiration or evaporation? Most often, the actual flux area is associated with canopy cover for trees and shrubs, and plant cover for herbaceous plants (fig. 8). Percentage of canopy or plant cover may vary with habitat type and resource availability. To accurately determine transpiration, the area of canopy cover or plant cover is required, as most fluxes are evaluated on canopy or plant cover. Note that for open water areas the fraction of plant coverage is equal to one. The fraction of plant coverage is defined as,

$$fCA(i, j, k) = \frac{\text{Canopy flux area for subgroup } k \text{ in polygon } j \text{ for cell } i}{\text{area of polygon } j \text{ for cell } i}.$$

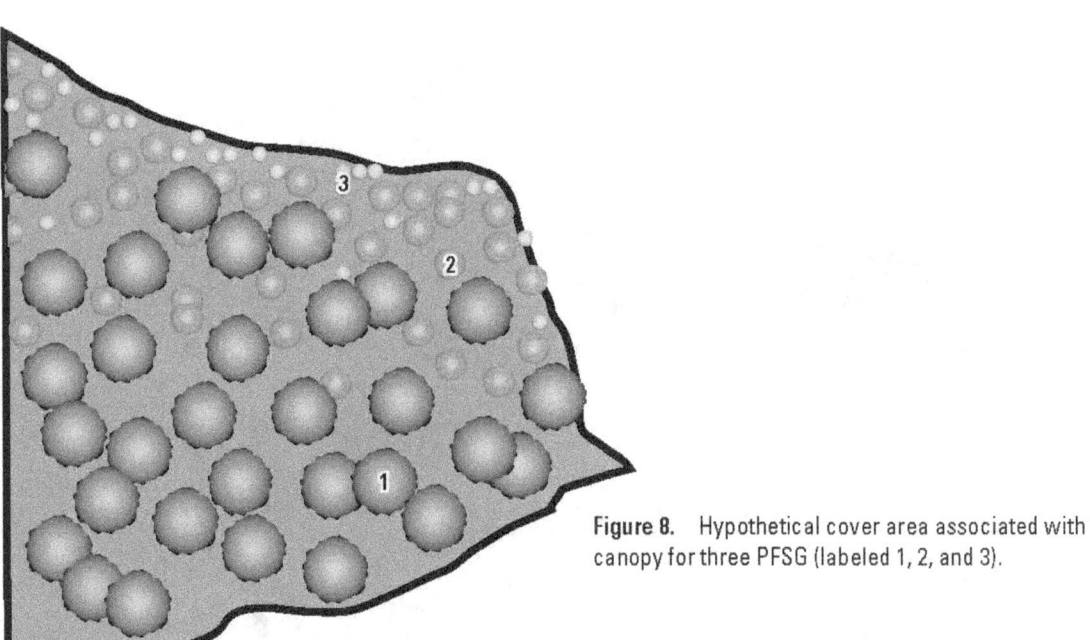

**Figure 8.**  Hypothetical cover area associated with canopy for three PFSG (labeled 1, 2, and 3).

## 8.0  Fractional Coverage

The fractional coverage of the plant functional subgroup $k$ in polygon $j$ for cell $i$ ($fCov(i,j,k)$) is given by the equation,

$$fCov(i, j, k) = fPA(i, j) \times fCA(i, j, k).$$

The components of fractional cover can be determined using a combination of GIS techniques, aerial photography and ground verification. The fractional covers ($fCov$) are input into RIP-ET for each plant functional subgroup in each polygon for each cell. These fractional covers may vary by stress periods. If a plant functional subgroup is not present in a polygon, its fraction is entered as zero.

## 9.0  Land-Surface Elevation

Within the model area, the land-surface elevation ($HSURF$) will vary from cell to cell. It also may vary within a cell. Because $HSURF$ is one of the parameters used to calculate extinction depths for the plant functional groups, the magnitude of the within-cell variability of $HSURF$ is important. A single value of $HSURF$ is assigned to each polygon. If terrain relief is small, the polygons may be large; however, where there is considerable terrain relief within cells (as in fig. 9), the user should consider using smaller polygons. The MODFLOW model assumes that the water-table elevation is uniform within the cell. Thus, extinction depths for the PFSGs are calculated based on the elevation of their associated polygons rather than the average elevation for the entire cell. The accuracy of $HSURF$ and thus the extinction depths are controlled by the user-determined polygon size and the resolution of the elevation data rather than by the model cell size.

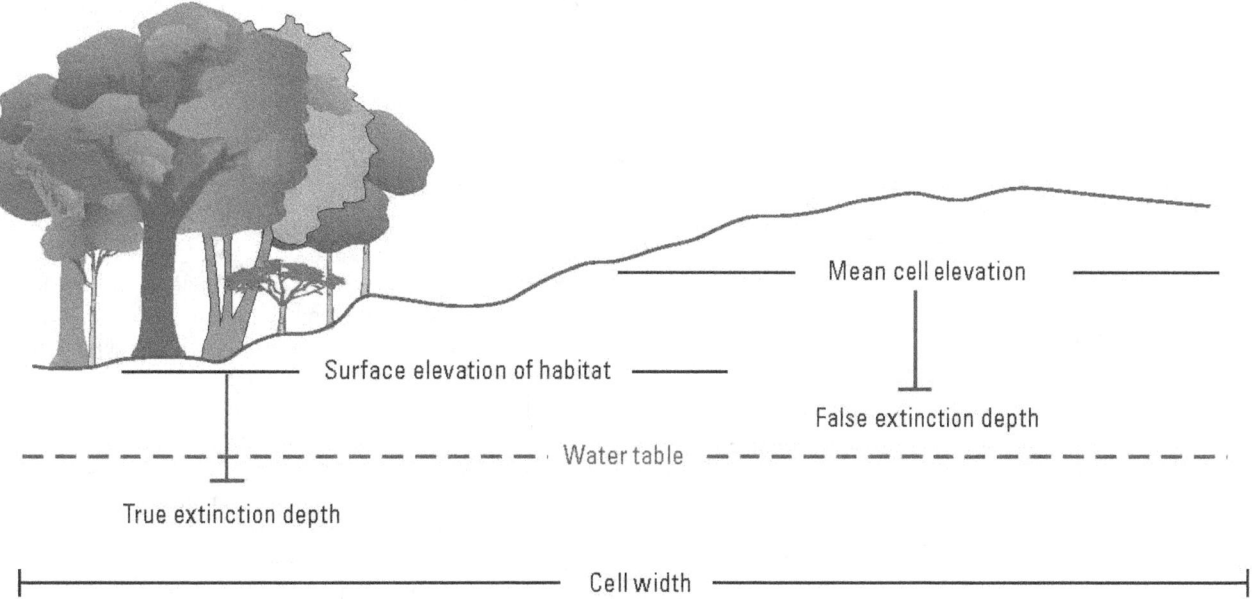

**Figure 9.**  Average surface elevation and actual habitat surface elevation.

# 10.0 Transpiration Calculations

Consider a plant functional subgroup present in a cell. If $(h(k), R(k))$ and $(h(k+1), R(k+1))$ are the coordinates of consecutive vertices that define the $k^{th}$ segment for that plant functional subgroup transpiration curve, and if $H$ is the hydraulic head in the cell, then the ET flux for the subgroup, $R(H)$ is given by

$$R(H) = R(k) + \frac{H - h(k)}{h(k+1) - h(k)} (R(k+1) - R(k))$$

for

$$h(k) \le H \le h(k+1),$$

(Hadley, 1964). The initial values are the extinction depth values (Hxd in fig. 3),

$$h(1) = \text{Hxd} = \text{HSURF} - \text{Sxd} - \text{Ard}$$

and

$$R(1) = 0.$$

The higher $k$ valued vertices are given by

$$h(k+1) = h(k) + \text{fdh}(k) \times \text{Ard}$$

and

$$R(k+1) = R(k) + \text{fdR}(k) \times \text{Rmax}.$$

To calculate the volumetric ET rate, $R(H)$ is multiplied by the cell area $[DELC \times DELR]$ and the fractional coverage, fCov, that is

$$ET(H) = R(H) \times [DELC \times DELR] \times \text{fCov}.$$

$R(H)$ can be rewritten,

$$R(H) = \frac{R(k+1) - R(k)}{h(k+1) - h(k)} H + \frac{R(k)h(k+1) - R(k+1)h(k)}{h(k+1) - h(k)}$$

for

$$h(k) \le H \le h(k+1).$$

A portion of $R(H)$ is dependent on the hydraulic head, $H$ (first term), while the other portion is independent of hydraulic head (second term). Define coefficient $C1$ as the head dependent term adjusted for area, and $C2$ as the head independent term, also adjusted for area,

$$C1 = \frac{R(k+1) - R(k)}{h(k+1) - h(k)} \times [DELC \times DELR] \times \text{fCov}$$

$$C2 = \frac{R(k)h(k+1) - R(k+1)h(k)}{h(k+1) - h(k)} \times [DELC \times DELR] \times \text{fCov},$$

both for $h(,k) \le H \le h(k+1)$. In the RIP-ET package subroutine GWF2RIP3FM, the variable $C1$ is added to the MODFLOW HCOF array, and $-C2$ (negative value) is added to the MODFLOW RHS array.

The above calculations are repeated in the Formulation-Procedure subroutine, GWF2RIP3FM, for each plant functional subgroup present in a cell, and for each cell designated as containing habitat. The budget module GWF2RIP3BD calculates, prints and stores the volumetric ET rate for a plant functional subgroup as

$$ET(H) = C1 \times H + C2.$$

The volumetric ET rate can be summed over plant functional subgroups to give the total volumetric ET rate within a cell, and further summed over the cells to give a global value of the volumetric ET rate within the modeled region.

# 11.0 Summary

Traditional approaches to modeling evapotranspiration processes within a groundwater model assume a piecewise linear relationship between the transpiration flux and hydraulic head (McDonald and Harbaugh, 1988; Harbaugh and McDonald, 1996a, 1996b; Banta, 2000). These approaches are limiting in several ways when a complex riparian system is to be simulated. First, a single curve is used to represent evaporation and all vegetation-type transpirations. However, transpiration fluxes vary between plant species or PFSG with groundwater depths due to differences in rooting depths and plant physiology, and differ significantly from evaporation rates. Second, no allowance is made for a decrease in transpiration caused by prolonged anoxic conditions. Third, current methodologies apply a single flux to the entire area of the designated cell, not allowing for fractional habitat coverage. Fourth, a single surface elevation is assigned to the entire cell despite the high surface variability characteristic of many riparian systems. The methodology of the RIP-ET package presented here replaces the traditional linear relation with a set of nonlinear dimensionless curves that reflects the ecophysiology of riparian and wetland plants.

In RIP-ET, the complex structure of wetland and riparian plant communities are simplified into a set of plant functional groups. Although the groups presented here are for semi-arid environments, the methodology can be applied universally. The five plant functional groups presented in this report are: obligate wetlands, shallow-rooted riparian, deep-rooted riparian, transitional riparian, and bare soil/open water. The groups can be further divided into subgroups based on plant size or density. Each subgroup has a transpiration flux curve that relates transpiration flux to water-table depth. A key characteristic of these flux curves is the existence of a saturated extinction depth along with the more traditional MODFLOW extinction depth, with the distance between them referred to as the active root depth. The transpiration flux curves are represented by a series of dimensionless, linear segments that are based on measured or approximated transpiration values.

RIP-ET requires a fractional coverage for each of the plant functional subgroups present within the groundwater model cell. A MODFLOW cell may contain several plant functional subgroups within its boundaries. Plant coverage within a cell is organized by polygons. A polygon consists of a single or mixture of plant functional subgroups and an approximately uniform land-surface elevation throughout. Typically, not all of a polygon contains habitat and only a portion of the habitat area has canopy or flux coverage. The fractional cover within a cell has two components: (1) the polygonal fraction of active habitat in a cell, and (2) fraction of plant flux area in a polygon. The product of PFSG's transpiration flux times the fractional plant coverage times the surface area of the cell gives the PFSG's transpiration rate for the cell. Summing over the PFSG in the cell gives the transpiration rate for that cell; summing over all the riparian cells gives the global transpiration rate for the modeled region. The combination of physically based transpiration curves and plant coverage estimates provides more accurate estimates of riparian transpiration and improves the modeling results.

Directly linking groundwater conditions and plant functional groups offers the opportunity to better simulate hydrologic/ecosystem interactions to help with improved management and restoration of these important ecosystems. By developing transpiration curves, this modeling approach allows for variability in vegetative conditions, the decrease in transpiration with elevated water tables, and separate representation of evaporation from transpiration. Furthermore, based on the types and coverage of plant functional subgroups present in the cell, multiple transpiration curves and surface elevations can be applied in a single model cell. In addition to the physical factors that govern evaporation, transpiration responds to a set of physiological and biometrical conditions. By decoupling the two processes, with the traditional linear curve being retained to model the evaporation, better estimates can be obtained for both. Furthermore, RIP-ET improves model accuracy by more effectively dealing with spatial issues of plant and water-table distribution. RIP-ET replaces the single-cell, single-transpiration-elevation value approach with multiple transpiration curves, and associated fractional coverages with variable surface elevations. Transpiration rates can now be calculated by determining the area of all habitat types present and then applying multiple transpiration curves to a single model cell.

# 12.0  Input Instructions for RIP-ET

Input to the RIP-ET is read from the file that has type "RIP" in the name file. All variables in datasets are read in free format if the option "FREE" is specified in the MODFLOW-2005 BASIC Package input file; otherwise, variables have 10-character fields, or are specified character field lengths.

## For Each Simulation

```
1. MAXRIP MAXPOLY IRIPCB IPHDRYRIPCB1
```

```
2. MAXTS MXSEG
```

Items 3, 4, and 5 are read for each plant functional subgroup.

```
3. RIPNM Sxd Ard Rmax Rsxd NuSeg
```

```
4.    fdh(1) fdh(2) ...  fdh(MXSEG)
```

```
5.    fdR(1) fdR(2) ...  fdR(MXSEG)
```

## For Each Stress Period

```
6. ITMP
```

The following data are read for each riparian cell.

```
7. Layer Row Column NPOLY
```

And for each polygon within the riparian cell, read.

```
8. HSURF fCov(1) ... fCov(MAXTS)
```

Repeat 7 and 8 until all riaparian cells have been read.

# 13.0  Explanation of Variables Read by the RIP-ET Package

| | |
|---|---|
| MAXRIP | is the maximum number of riparian cells. |
| MAXPOLY | is the maximum number of polygons in a cell. |
| IRIPCB | is a cell-by-cell flow flag and unit number.<br>IRIPCB>0, unit number to which the total evapotranspiration rate for each cell will be written when SAVE BUDGET or a non-zero value of ICBC is specified in output control.<br>IRIPCB=0, cell-by-cell terms not written.<br>IRIPCB<0, the ET rate for each plant functional subgroup will be written to the LIST file by cell when SAVE BUDGET or a non-zero value of ICBCFL is specified in output control. |
| IRIPCB1 | is a flag and unit number.<br>IRIPCB1>0, unit number to which the location, land-surface elevation, and transpiration rates for each plant functional subgroup and cell are saved.<br>IRIPCB1≤0, the values are not saved. |
| MAXTS | is the total number of plant functional subgroups. At present, MAXTS should less than or equal to 20. |
| MXSEG | is the maximum number of segments in any plant functional subgroup for the interpolation of transpiration canopy flux<br>as a function of hydraulic head. |
| RIPNM | are the plant functional subgroup names, such as:<br>Obligate wetland: low, medium or high density.<br>Shallow-rooted riparian: low, medium, or high density.<br>Deep-rooted riparian: small, medium, or large size.<br>Transitional-riparian: small, medium, or large size.<br>RIPNM is string of up to 24 characters. |
| Sxd | is the saturated extinction depth with respect to the surface elevation, (L). (A negative depth is above land surface). |
| Ard | is the active root depth, (L). |
| Rmax | is the maximum transpiration or evaporation flux, (L/t). |
| Rsxd | is the transpiration canopy flux at the saturated extinction depth or the maximum evaporation rate, (L/t). |
| NuSeg | is the number of active segments to perform a linear interpolation for a plant functional subgroup. |
| fdh | is the dimensionless active root depth segment (eqn. 5.1, dimensionless). |
| fdR | is the dimensionless flux segment (eqn. 5.2, dimensionless). |
| ITMP | is a flag or the number of riparian cells.<br>≥ 0, number of riparian cells active during the current stress period.<br>< 0, same riparian cells active during last stress period will be active during the current stress period. |
| Layer | is the layer number of the riparian cell. |
| Row | is the row number of the riparian cell. |
| Column | is the column number of the riparian cell. |
| NPOLY | is the number of polygons in a cell. |
| HSURF | is the land-surface elevation of a polygon in a riparian cell. |
| fCov | is the fraction of land coverage of a plant functional subgroup for a polygon within a cell; for each polygon there may be multiple plant functional subgroups. |

# 14.0 Example

A hypothetical region, Dry Alkaline Valley, extends over 200 mi² and is bounded to the north and south by mountain ranges that act as no-flow boundaries. The basin is underlain by a single unconfined aquifer with the hydraulic conductivity distribution and pertinent geometry given in Exmpl.bcf (datasets are shown in Section 14.1). A large lake to the northwest behaves hydrologically as a prescribed head boundary, and is the source of the river that transects the basin from west to east. The groundwater in the basin and the river both discharge to the east. Riparian habitats border the river (see fig. 10). Dry Alkaline Valley has two seasons: a growing season from April to September and a dormant season from October to March. The stream inflow from the lake is assumed to be the same in both seasons. Stream data are given in Exmpl.str. The outflow from the eastern boundary is simulated as wells and is assumed to be the same for both seasons; its data are given in Exmpl.wel. Output control and the SIP solution packages are specified in Exmpl.oc and Exmpl.sip, respectively. The riparian evapotranspiration dataset is called Exmpl.rip.

The Dry Alkaline Valley habitat consists of a mixture of small, medium, and large deep-rooted riparian subgroups distributed over two terraces mirrored on each side of the stream channel (figs. 11 and 12). The inner and outer terraces are 100 and 200 ft wide, respectively. The stream channel is 100 ft wide.

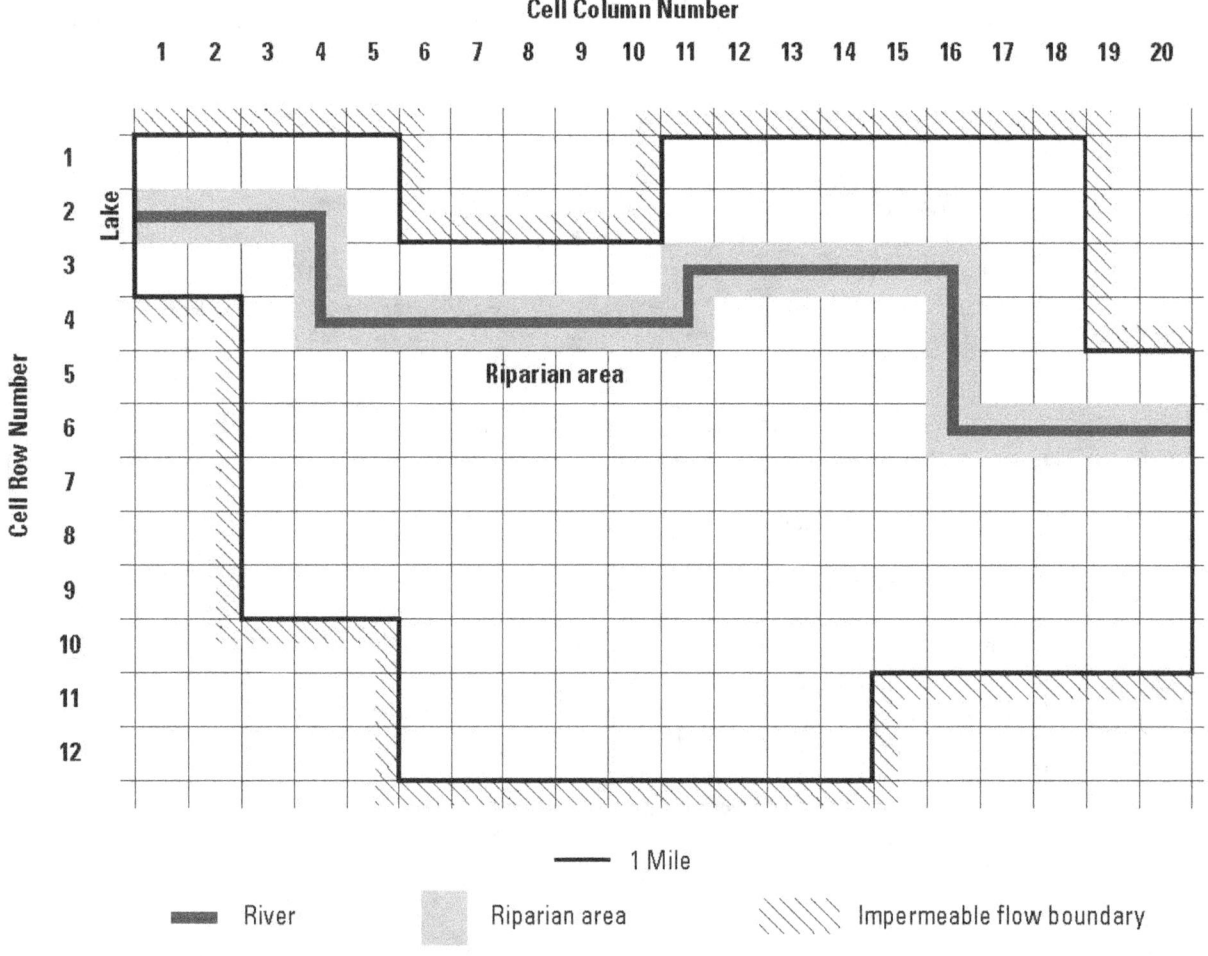

**Figure 10.** Aerial view of Dry Alkaline Valley. (The riparian area and stream system are magnified for ease of viewing cell location.)

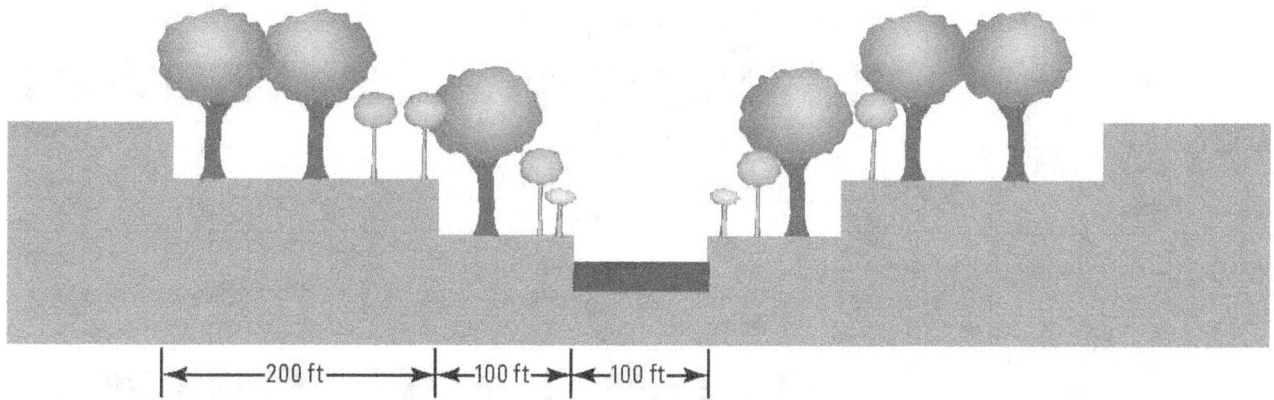

**Figure 11.**   Cross-section of Dry Alkaline Valley riparian area.

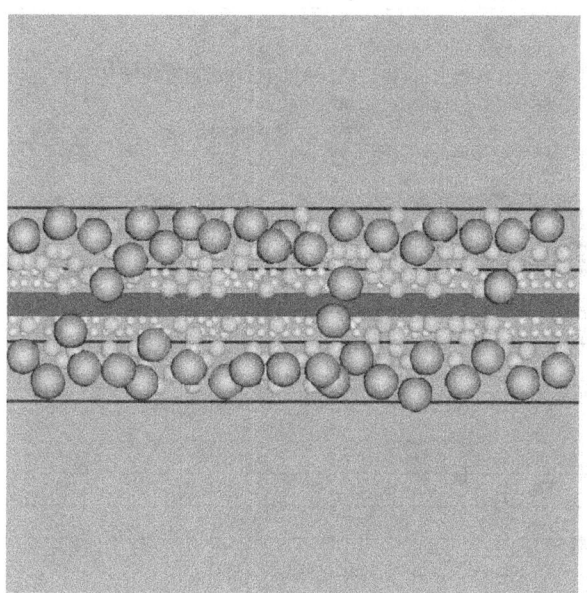

**Figure 12.**   Canopy coverage for typical riparian cell. (Note: The area of the cell and the area of the stream and riparian area are not in proper proportions in order to aid visualization).

Each riparian cell has four polygons: two on each side of the river. Within the polygons of a cell, the land-surface elevation has only small changes and is considered constant. Table 2 presents the polygon areas and fractions of polygon

**Table 2.**   Area and fraction of inner and outer terraces.

[ft², square foot]

| Polygon | Area (ft²) | fPA |
|---|---|---|
| Terrace outer left | 1,056,000 | 0.03788 |
| Terrace inner left | 528,000 | 0.01894 |
| Terrace inner right | 528,000 | 0.01894 |
| Terrace outer right | 1,056,000 | 0.03788 |

area to cell areal ($fPA(i,j)$) for polygon $j$, cell $i$). All cells are assumed to have the same polygon area and the same cell surface area. Traversing the river from the lake to the outflow to the east, the terraces are labeled: Terrace Outer Left, Terrace Inner Left, Terrace Inner Right, and Terrace Outer Right. The cell surface area is 1 mi² (27,878,400 ft²).

The outer terrace areas left and right comprised approximately 40 percent bare-ground flux and 60 percent canopy flux. For the canopy flux area in the outer terraces, 60 percent are large trees, and 40 percent are medium trees. The inner terrace areas are composed of 33 percent bare ground and 67 percent canopy. The canopy flux area has 40.5 percent small trees, 35.5 percent medium trees, and 24 percent large trees. Note that for the example, the percentages sum to 100 percent. The sum may be higher than 100 percent if the large trees shelter the smaller ones or if additional bare-ground evaporation is included under the tree canopy.

The resulting coverage breakdowns for each plant functional group are listed in tables 3 and 4. During the dormant season, the terrace polygons are modeled as bare ground.

The HSURF for the inner and outer terraces are 5 and 10 ft, respectively, above the top of the streambed.

During the first season (April–September), the habitat is active. Maximum transpiration fluxes (Rmax) are 0.2003 cm/d ($7.61 \times 10^{-8}$ ft/s) for small trees, 0.2615 cm/d ($9.91 \times 10^{-8}$ ft/s) for medium trees, and 0.2944 cm/d ($1.12 \times 10^{-7}$ ft/s) for large trees (table 5). The active root depths (Ard) are 500 cm (16.40 ft) for the medium and large trees, and 400 cm (13.12 ft) for small trees (fig. 13). The saturated extinction depth (Sxd) is at land surface for all subgroups and is thus zero. The transpiration flux at the saturated extinction depth (Rsxd) is zero for all subgroups. The fdh and fdR values calculated for the above curves are presented in table 6. Recall that the fdh and fdR are dimensionless (eqns. 5.1 and 5.2). Because the other MODFLOW data have units of feet and seconds, the values of Ard, Sxd, and Rmax also must be in feet and feet per second.

The evaporation flux curve for the bare ground/open water polygon area during the first season and for the entire terrace polygon areas the second season, follows the traditional MODFLOW form as shown in figure 14. The maximum evaporation flux (Rmax) is assumed to be 0.550 cm/d ($2.09 \times 10^{-7}$ ft/s), with an extinction depth of 100 cm (3.28 ft). The maximum ET surface coincides with the head at saturated extinction depth (Hsxd) and is assumed to be the land surface of the terraces. The fdh and the fdR values for the evaporation flux curves are both equal to 1.0.

**Table 3.** Fractional coverage for plant functional subgroups per model cell on inner terraces.

| PFSG type | fPA | fCA | fCov |
|---|---|---|---|
| Deep-rooted: Small | 0.01894 | 0.67*0.405 = 0.271 | 0.01894*0.271 = 0.00513 |
| Deep-rooted: Medium | 0.01894 | 0.67*0.355 = 0.238 | 0.01894*0.238 = 0.00451 |
| Deep-rooted: Large | 0.01894 | 0.67*0.240 = 0.161 | 0.01894*0.161 = 0.00305 |
| Bare ground | 0.01894 | 0.33*1.00 = 0.330 | 0.01894*0.330 = 0.00625 |

**Table 4.** Fractional coverage for plant functional subgroups per model cell on outer terraces.

| PFSG type | fPA | fCA | fCov |
|---|---|---|---|
| Deep-rooted: Medium | 0.03788 | 0.60*0.40 = 0.24 | 0.03788*0.24 = 0.00909 |
| Deep-rooted: Large | 0.03788 | 0.60*0.60 = 0.36 | 0.03788*0.36 = 0.01364 |
| Bare ground | 0.03788 | 0.40*1.00 = 0.40 | 0.03788*0.40 = 0.01515 |

**Table 5.** Vertices for the transpiration flux functions for the deep-rooted riparian subgroups.

[cm, centimeter; cm/d, centimeter per day]

| Depth (cm) | Flux (cm/d) | | |
|---|---|---|---|
| | Small | Medium | Large |
| 0.00 | 0.0000 | 0.0000 | 0.0000 |
| −75.00 | 0.1742 | 0.1506 | 0.1400 |
| −100.00 | 0.2003 | 0.2357 | 0.2653 |
| −150.00 | 0.1602 | 0.2615 | 0.2944 |
| −200.00 | 0.1001 | 0.2450 | 0.2759 |
| −300.00 | 0.0501 | 0.1650 | 0.2064 |
| −400.00 | 0.0000 | 0.0838 | 0.1032 |
| −500.00 | 0.0000 | 0.0000 | 0.0000 |

**Table 6.** Dimensionless fdh and fdR for deep-rooted riparian subgroups.

[Counting starts at extinction depth and proceeds toward land surface]

| Small | | Medium | | Large | |
|---|---|---|---|---|---|
| fdh | fdR | fdh | fdR | fdh | fdR |
| 0.25 | 0.25012 | 0.20 | 0.32184 | 0.20 | 0.35054 |
| 0.25 | 0.24963 | 0.20 | 0.31034 | 0.20 | 0.35054 |
| 0.125 | 0.30005 | 0.20 | 0.30651 | 0.20 | 0.23607 |
| 0.125 | 0.20020 | 0.10 | 0.06130 | 0.10 | 0.06284 |
| 0.0625 | −0.13030 | 0.10 | −0.09579 | 0.10 | −0.09885 |
| 0.1875 | −0.86970 | 0.05 | −0.32567 | 0.05 | −0.42561 |
| | | 0.15 | −0.57854 | 0.15 | −0.47554 |

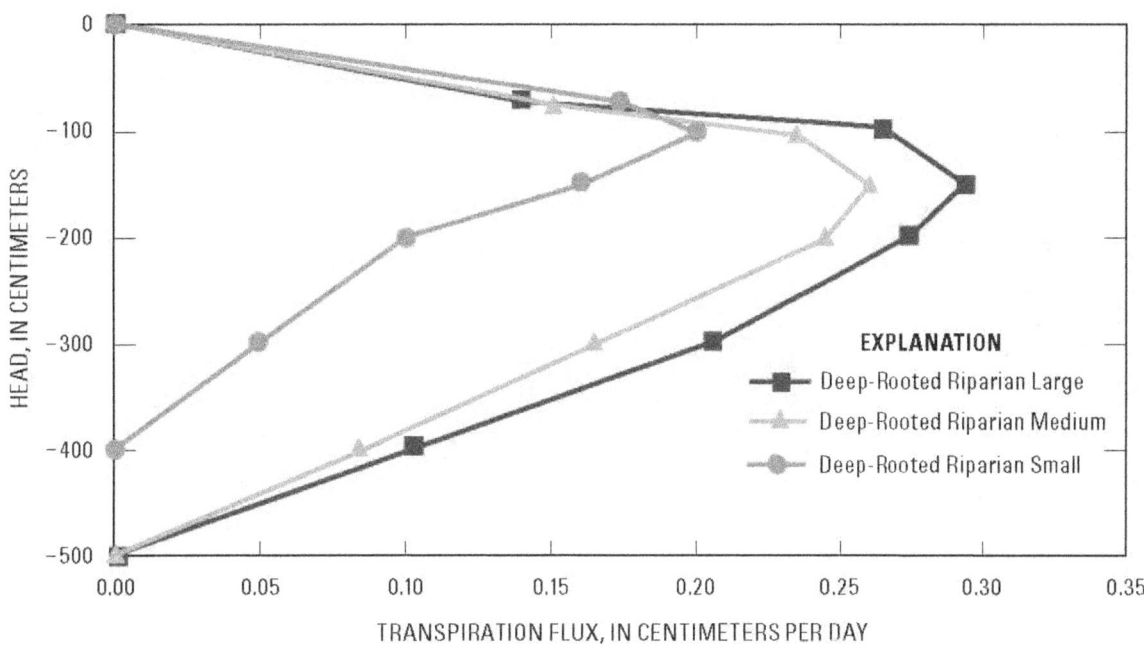

**Figure 13.** Seasonal canopy-area transpiration curves for the three size classes of deep-rooted riparian subgroups (cottonwood and willow).

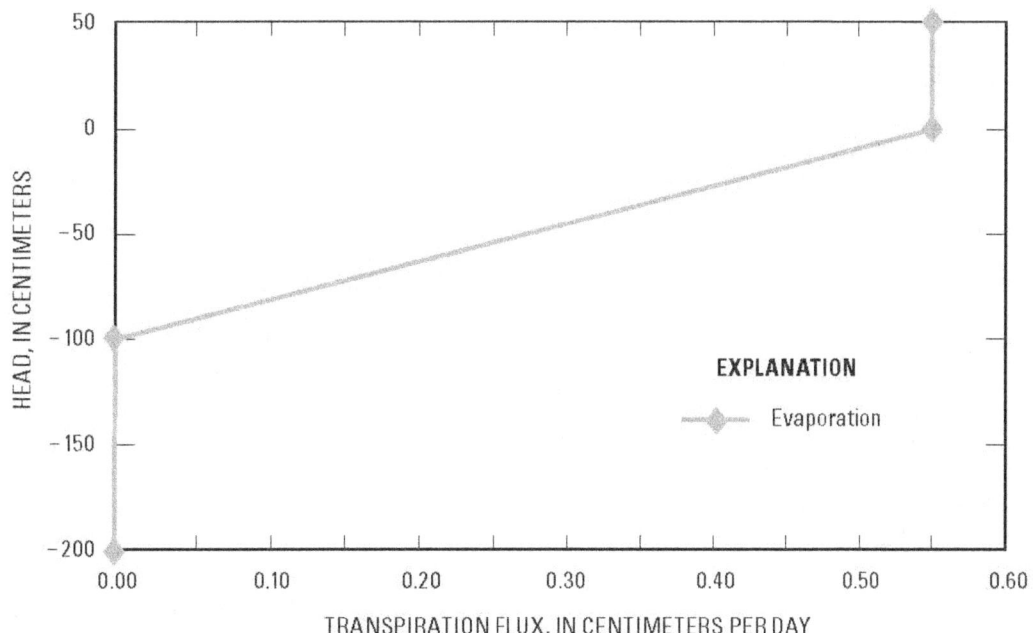

**Figure 14.** Traditional MODFLOW flux curve for evaporation.

## 14.1 Datasets for Example

The following datasets for MODFLOW-2005 all use the FREE format. The length unit is feet, and the time unit is seconds.

## Name File: Exmpl.nam

```
LIST    26    Exmpl.lst
BAS6    1     Exmpl.ba1
BCF6    11    Exmpl.bcf
WEL     12    Exmpl.wel
STR     14    Exmpl.str
SIP     16    Exmpl.sip
OC      17    Exmpl.oc
RIP     20    Exmpl3.rip
DIS     21    Exmpl.dis
DATA    30    Exmpl.hed
DATA    40    Initial.hed
DATA    45    LandSurf.dem
DATA    50    Exmpl.ret
```

## DIS File: Exmpl.dis

```
# Riparian ET Test Model, 2 Season Model
         1         12        20         2         1         1
         0
 CONSTANT  5280.0000                        DELR
 CONSTANT  5280.0000                        DELC
 EXTERNAL  45   1.0   (10F6.0)    7         TOP
 CONSTANT  3500                             BOT
15778800.          1 1.0000000   TR
15778800.          1 1.0000000   TR
```

## BAS File: Exmpl.ba1

```
# Riparian ET Test Model, 2 Season Model
  FREE
INTERNAL   1                    (20I3)         3
 -1  1  1  1  1  0  0  0  0  0  1  1  1  1  1  1  1  1  0  0
 -1  1  1  1  1  0  0  0  0  0  1  1  1  1  1  1  1  1  0  0
 -1  1  1  1  1  1  1  1  1  1  1  1  1  1  1  1  1  1  0  0
  0  0  1  1  1  1  1  1  1  1  1  1  1  1  1  1  1  1  0  0
  0  0  1  1  1  1  1  1  1  1  1  1  1  1  1  1  1  1  1  1
  0  0  1  1  1  1  1  1  1  1  1  1  1  1  1  1  1  1  1  1
  0  0  1  1  1  1  1  1  1  1  1  1  1  1  1  1  1  1  1  1
  0  0  1  1  1  1  1  1  1  1  1  1  1  1  1  1  1  1  1  1
  0  0  1  1  1  1  1  1  1  1  1  1  1  1  1  1  1  1  1  1
  0  0  0  0  0  1  1  1  1  1  1  1  1  1  1  1  1  1  1  1
  0  0  0  0  0  1  1  1  1  1  1  1  1  1  0  0  0  0  0  0
  0  0  0  0  0  1  1  1  1  1  1  1  1  1  0  0  0  0  0  0
 -999.0000
EXTERNAL  40   1.0   (10F8.1)    3
```

## BCF File: Exmpl.bcf

```
       -1    1.0E+30    0     0.0    0     0
 1
 CONSTANT  1.0000000                        TRPY
 CONSTANT  0.01                             Specific yield
 CONSTANT  0.001050                         Hydraulic conductivity
```

STR File: Exmp1.str

|    | 25 |    | 1  |    | 0    |      | 0     | 1      | 1.486 | -1    | 0     |
|----|----|----|----|----|------|------|-------|--------|-------|-------|-------|
|    | 25 |    | 0  |    | 0    |      |       |        |       |       |       |
| 1  | 2  | 2  | 1  | 1  |      | 100. | 3796. | .0924  | 3776. | 3786. |
| 1  | 2  | 3  | 1  | 2  |      | 0.   | 0.    | .0924  | 3773. | 3783. |
| 1  | 2  | 4  | 1  | 3  |      | 0.   | 0.    | .0924  | 3769. | 3779. |
| 1  | 3  | 4  | 1  | 4  |      | 0.   | 0.    | .0924  | 3768. | 3778. |
| 1  | 4  | 4  | 1  | 5  |      | 0.   | 0.    | .0924  | 3767. | 3777. |
| 1  | 4  | 5  | 1  | 6  |      | 0.   | 0.    | .0924  | 3765. | 3775. |
| 1  | 4  | 6  | 1  | 7  |      | 0.   | 0.    | .0924  | 3763. | 3773. |
| 1  | 4  | 7  | 1  | 8  |      | 0.   | 0.    | .0924  | 3758. | 3768. |
| 1  | 4  | 8  | 1  | 9  |      | 0.   | 0.    | .0924  | 3754. | 3764. |
| 1  | 4  | 9  | 1  | 10 |      | 0.   | 0.    | .0924  | 3750. | 3760. |
| 1  | 4  | 10 | 1  | 11 |      | 0.   | 0.    | .0924  | 3747. | 3757. |
| 1  | 4  | 11 | 1  | 12 |      | 0.   | 0.    | .0924  | 3743. | 3753. |
| 1  | 3  | 11 | 1  | 13 |      | 0.   | 0.    | .0924  | 3742. | 3752. |
| 1  | 3  | 12 | 1  | 14 |      | 0.   | 0.    | .0924  | 3739. | 3749. |
| 1  | 3  | 13 | 1  | 15 |      | 0.   | 0.    | .0924  | 3736. | 3746. |
| 1  | 3  | 14 | 1  | 16 |      | 0.   | 0.    | .0924  | 3732. | 3742. |
| 1  | 3  | 15 | 1  | 17 |      | 0.   | 0.    | .0924  | 3728. | 3738. |
| 1  | 3  | 16 | 1  | 18 |      | 0.   | 0.    | .0924  | 3727. | 3737. |
| 1  | 4  | 16 | 1  | 19 |      | 0.   | 0.    | .0924  | 3726. | 3736. |
| 1  | 5  | 16 | 1  | 20 |      | 0.   | 0.    | .0924  | 3725. | 3735. |
| 1  | 6  | 16 | 1  | 21 |      | 0.   | 0.    | .0924  | 3724. | 3734. |
| 1  | 6  | 17 | 1  | 22 |      | 0.   | 0.    | .0924  | 3721. | 3731. |
| 1  | 6  | 18 | 1  | 23 |      | 0.   | 0.    | .0924  | 3717. | 3727. |
| 1  | 6  | 19 | 1  | 24 |      | 0.   | 0.    | .0924  | 3714. | 3724. |
| 1  | 6  | 20 | 1  | 25 |      | 0.   | 0.    | .0924  | 3710. | 3720. |

| 100. | .000568 | .03 |
|------|---------|-----|
| 100. | .000758 | .03 |
| 100. | .000189 | .03 |
| 100. | .000189 | .03 |
| 100. | .000379 | .03 |
| 100. | .000379 | .03 |
| 100. | .000947 | .03 |
| 100. | .000758 | .03 |
| 100. | .000758 | .03 |
| 100. | .000568 | .03 |
| 100. | .000758 | .03 |
| 100. | .000189 | .03 |
| 100. | .000189 | .03 |
| 100. | .000568 | .03 |
| 100. | .000568 | .03 |
| 100. | .000758 | .03 |
| 100. | .000758 | .03 |
| 100. | .000189 | .03 |
| 100. | .000189 | .03 |
| 100. | .000189 | .03 |
| 100. | .000189 | .03 |
| 100. | .000568 | .03 |
| 100. | .000758 | .03 |
| 100. | .000568 | .03 |
| 100. | .000758 | .03 |
| -1   | 0       | 0   |

## WEL File: Exmp1.wel

```
6          -1
1
1          5         20 -1.0
1          6         20 -1.0
1          7         20 -1.0
1          8         20 -1.0
1          9         20 -1.0
1         10         20 -1.0
-1
```

## SIP File: Exmp1.sip

```
    100
1.0000000 0.0001000     1    0.0     0
```

## OC File: Exmp1.oc

```
HEAD     PRINT   FORMAT 7
HEAD     SAVE    FORMAT (10F8.1)
HEAD     SAVE    UNIT 30
COMPACT BUDGET
PERIOD 1 STEP 1
PRINT HEAD
PRINT BUDGET
SAVE BUDGET
SAVE HEAD
PERIOD 2 STEP 1
PRINT HEAD
PRINT BUDGET
SAVE BUDGET
SAVE HEAD
```

## Initial Head File: Initial.hed

```
3800.0   3793.5   3788.0   3784.0   3781.9   -999.0   -999.0   -999.0   -999.0   -999.0
3746.3   3745.3   3743.5   3741.5   3739.5   3737.7   3736.3   3735.6   -999.0   -999.0
3800.0   3792.5   3786.4   3782.1   3779.8   -999.0   -999.0   -999.0   -999.0   -999.0
3747.4   3745.9   3743.8   3741.5   3739.3   3737.4   3735.7   3734.8   -999.0   -999.0
3800.0   3792.2   3783.8   3779.2   3775.4   3769.9   3765.6   3761.6   3757.8   3754.1
3750.1   3747.1   3744.3   3741.5   3738.9   3736.7   3734.5   3733.1   -999.0   -999.0
-999.0   -999.0   3777.2   3775.3   3772.5   3768.9   3765.1   3761.4   3757.8   3754.3
3750.7   3747.0   3743.9   3740.9   3738.0   3735.4   3732.4   3730.0   -999.0   -999.0
-999.0   -999.0   3772.6   3771.3   3769.1   3766.3   3763.1   3759.9   3756.6   3753.3
3749.9   3746.5   3743.3   3740.1   3736.9   3733.7   3729.6   3724.5   3715.8   3707.0
-999.0   -999.0   3769.1   3768.1   3766.4   3764.0   3761.3   3758.4   3755.4   3752.3
3749.2   3745.9   3742.6   3739.2   3735.7   3732.1   3727.7   3722.3   3715.4   3707.3
-999.0   -999.0   3766.6   3765.8   3764.2   3762.0   3759.6   3757.0   3754.3   3751.4
3748.4   3745.3   3742.0   3738.4   3734.5   3730.2   3725.2   3719.3   3712.2   3703.9
-999.0   -999.0   3764.9   3764.1   3762.5   3760.3   3758.1   3755.8   3753.3   3750.7
3747.9   3744.8   3741.6   3737.9   3733.6   3728.8   3723.4   3717.1   3709.8   3701.4
-999.0   -999.0   3764.1   3763.2   3761.5   3758.7   3756.6   3754.6   3752.4   3750.0
3747.4   3744.6   3741.5   3737.8   3733.2   3728.0   3722.3   3715.8   3708.3   3699.8
-999.0   -999.0   -999.0   -999.0   -999.0   3756.2   3755.1   3753.6   3751.7   3749.6
3747.2   3744.7   3742.0   3738.6   3733.2   3727.7   3721.8   3715.1   3707.6   3699.0
-999.0   -999.0   -999.0   -999.0   -999.0   3754.8   3754.1   3752.8   3751.2   3749.3
3747.2   3745.1   3743.0   3741.4   -999.0   -999.0   -999.0   -999.0   -999.0   -999.0
-999.0   -999.0   -999.0   -999.0   -999.0   3754.2   3753.6   3752.4   3750.9   3749.1
3747.2   3745.3   3743.6   3742.5   -999.0   -999.0   -999.0   -999.0   -999.0   -999.0
```

## Land-Surface Elevation File: LandSurf.dem

```
3800.0   3799.0   3789.0   3787.0   3783.8   -999.0   -999.0   -999.0   -999.0   -999.0
3760.1   3755.9   3755.7   3748.6   3744.3   3742.9   3740.4   3739.0   -999.0   -999.0
3800.0   3794.0   3787.5   3782.5   3783.6   -999.0   -999.0   -999.0   -999.0   -999.0
3757.4   3758.5   3752.6   3746.9   3742.7   3741.7   3740.1   3738.3   -999.0   -999.0
3800.0   3798.0   3788.0   3782.5   3781.6   3778.3   3774.4   3771.7   3767.9   3764.3
3756.5   3753.5   3750.5   3746.5   3742.0   3741.0   3738.6   3735.8   -999.0   -999.0
-999.0   -999.0   3788.9   3781.0   3779.0   3777.5   3772.5   3768.0   3764.0   3761.0
3757.5   3755.7   3752.7   3747.6   3744.4   3740.0   3738.1   3734.2   -999.0   -999.0
-999.0   -999.0   3790.2   3782.4   3782.1   3779.5   3772.5   3768.6   3767.1   3762.8
3760.0   3758.8   3753.2   3749.4   3745.6   3739.0   3736.6   3733.4   3727.5   3724.3
-999.0   -999.0   3792.6   3782.5   3785.8   3780.1   3776.5   3771.5   3767.8   3766.3
3761.9   3758.9   3755.2   3749.8   3747.6   3738.0   3735.0   3730.5   3727.5   3723.5
-999.0   -999.0   3792.7   3786.3   3789.6   3783.5   3779.1   3773.9   3769.5   3768.4
3764.2   3760.8   3757.0   3752.3   3749.1   3738.3   3737.8   3731.9   3727.9   3724.1
-999.0   -999.0   3794.0   3786.8   3790.3   3787.4   3779.7   3774.5   3770.7   3772.2
3765.6   3763.1   3758.7   3755.2   3751.8   3738.9   3740.4   3731.9   3729.6   3724.4
-999.0   -999.0   3796.1   3789.5   3793.7   3790.4   3783.1   3774.9   3773.5   3773.3
3768.7   3763.8   3760.5   3755.7   3753.7   3739.7   3742.9   3733.5   3730.0   3725.2
-999.0   -999.0   -999.0   -999.0   -999.0   3793.6   3786.9   3776.1   3777.0   3774.2
3772.3   3766.8   3761.7   3756.0   3754.7   3740.4   3744.8   3734.6   3731.4   3726.1
-999.0   -999.0   -999.0   -999.0   -999.0   3794.6   3788.6   3776.7   3778.7   3778.2
3772.7   3767.0   3763.7   3756.7   -999.0   -999.0   -999.0   -999.0   -999.0   -999.0
-999.0   -999.0   -999.0   -999.0   -999.0   3797.8   3792.1   3780.6   3780.9   3778.9
3775.2   3767.2   3766.6   3758.6   -999.0   -999.0   -999.0   -999.0   -999.0   -999.0
```

## 14.2  Riparian Dataset

The dataset (shaded for delineation) for Exmpl.rip is as follows,

```
        25          4          -1          50
         4          7
"D.R. Riparian Small     "      0.00      13.12 0.761E-07 0.000E+00                6
    0.25000     0.25000     0.12500     0.12500     0.06250     0.18750
    0.25012     0.24963     0.30005     0.20020    -0.13030    -0.86970
"D.R. Riparian Medium    "      0.00      16.40 0.991E-07 0.000E+00                7
    0.20000     0.20000     0.20000     0.10000     0.10000     0.05000     0.15000
    0.32184     0.31034     0.30651     0.06130    -0.09579    -0.32567    -0.57854
"D.R. Riparian Large     "      0.00      16.40 0.112E-06 0.000E+00                7
    0.20000     0.20000     0.20000     0.10000     0.10000     0.05000     0.15000
    0.35054     0.35054     0.23607     0.06284    -0.09885    -0.42561    -0.47554
"Evaporation             "      0.00       3.28 0.209E-06 0.209E-06                1
    1.00000
    1.00000
        25
         1          2          2          4
    3796.00     0.00000     0.00909     0.01364     0.01515
    3791.00     0.00513     0.00451     0.00305     0.00625
    3791.00     0.00513     0.00451     0.00305     0.00625
    3796.00     0.00000     0.00909     0.01364     0.01515
         1          2          3          4
    3793.00     0.00000     0.00909     0.01364     0.01515
    3788.00     0.00513     0.00451     0.00305     0.00625
    3788.00     0.00513     0.00451     0.00305     0.00625
    3793.00     0.00000     0.00909     0.01364     0.01515
         1          2          4          4
    3789.00     0.00000     0.00909     0.01364     0.01515
    3784.00     0.00513     0.00451     0.00305     0.00625
    3784.00     0.00513     0.00451     0.00305     0.00625
    3789.00     0.00000     0.00909     0.01364     0.01515
         1          3          4          4
    3788.00     0.00000     0.00909     0.01364     0.01515
    3783.00     0.00513     0.00451     0.00305     0.00625
    3783.00     0.00513     0.00451     0.00305     0.00625
    3788.00     0.00000     0.00909     0.01364     0.01515
         1          4          4          4
    3787.00     0.00000     0.00909     0.01364     0.01515
    3782.00     0.00513     0.00451     0.00305     0.00625
    3782.00     0.00513     0.00451     0.00305     0.00625
    3787.00     0.00000     0.00909     0.01364     0.01515
         1          4          5          4
    3785.00     0.00000     0.00909     0.01364     0.01515
    3780.00     0.00513     0.00451     0.00305     0.00625
    3780.00     0.00513     0.00451     0.00305     0.00625
    3785.00     0.00000     0.00909     0.01364     0.01515
         1          4          6          4
    3783.00     0.00000     0.00909     0.01364     0.01515
    3778.00     0.00513     0.00451     0.00305     0.00625
    3778.00     0.00513     0.00451     0.00305     0.00625
    3783.00     0.00000     0.00909     0.01364     0.01515
         1          4          7          4
    3778.00     0.00000     0.00909     0.01364     0.01515
    3773.00     0.00513     0.00451     0.00305     0.00625
    3773.00     0.00513     0.00451     0.00305     0.00625
    3778.00     0.00000     0.00909     0.01364     0.01515
         1          4          8          4
    3774.00     0.00000     0.00909     0.01364     0.01515
    3769.00     0.00513     0.00451     0.00305     0.00625
    3769.00     0.00513     0.00451     0.00305     0.00625
    3774.00     0.00000     0.00909     0.01364     0.01515
         1          4          9          4
    3770.00     0.00000     0.00909     0.01364     0.01515
    3765.00     0.00513     0.00451     0.00305     0.00625
    3765.00     0.00513     0.00451     0.00305     0.00625
    3770.00     0.00000     0.00909     0.01364     0.01515
```

| 1 | 4 | 10 | 4 | |
|---|---|---|---|---|
| 3767.00 | 0.00000 | 0.00909 | 0.01364 | 0.01515 |
| 3762.00 | 0.00513 | 0.00451 | 0.00305 | 0.00625 |
| 3762.00 | 0.00513 | 0.00451 | 0.00305 | 0.00625 |
| 3767.00 | 0.00000 | 0.00909 | 0.01364 | 0.01515 |
| 1 | 4 | 11 | 4 | |
| 3763.00 | 0.00000 | 0.00909 | 0.01364 | 0.01515 |
| 3758.00 | 0.00513 | 0.00451 | 0.00305 | 0.00625 |
| 3758.00 | 0.00513 | 0.00451 | 0.00305 | 0.00625 |
| 3763.00 | 0.00000 | 0.00909 | 0.01364 | 0.01515 |
| 1 | 3 | 11 | 4 | |
| 3762.00 | 0.00000 | 0.00909 | 0.01364 | 0.01515 |
| 3757.00 | 0.00513 | 0.00451 | 0.00305 | 0.00625 |
| 3757.00 | 0.00513 | 0.00451 | 0.00305 | 0.00625 |
| 3762.00 | 0.00000 | 0.00909 | 0.01364 | 0.01515 |
| 1 | 3 | 12 | 4 | |
| 3759.00 | 0.00000 | 0.00909 | 0.01364 | 0.01515 |
| 3754.00 | 0.00513 | 0.00451 | 0.00305 | 0.00625 |
| 3754.00 | 0.00513 | 0.00451 | 0.00305 | 0.00625 |
| 3759.00 | 0.00000 | 0.00909 | 0.01364 | 0.01515 |
| 1 | 3 | 13 | 4 | |
| 3756.00 | 0.00000 | 0.00909 | 0.01364 | 0.01515 |
| 3751.00 | 0.00513 | 0.00451 | 0.00305 | 0.00625 |
| 3751.00 | 0.00513 | 0.00451 | 0.00305 | 0.00625 |
| 3756.00 | 0.00000 | 0.00909 | 0.01364 | 0.01515 |
| 1 | 3 | 14 | 4 | |
| 3752.00 | 0.00000 | 0.00909 | 0.01364 | 0.01515 |
| 3747.00 | 0.00513 | 0.00451 | 0.00305 | 0.00625 |
| 3747.00 | 0.00513 | 0.00451 | 0.00305 | 0.00625 |
| 3752.00 | 0.00000 | 0.00909 | 0.01364 | 0.01515 |
| 1 | 3 | 15 | 4 | |
| 3748.00 | 0.00000 | 0.00909 | 0.01364 | 0.01515 |
| 3743.00 | 0.00513 | 0.00451 | 0.00305 | 0.00625 |
| 3743.00 | 0.00513 | 0.00451 | 0.00305 | 0.00625 |
| 3748.00 | 0.00000 | 0.00909 | 0.01364 | 0.01515 |
| 1 | 3 | 16 | 4 | |
| 3747.00 | 0.00000 | 0.00909 | 0.01364 | 0.01515 |
| 3742.00 | 0.00513 | 0.00451 | 0.00305 | 0.00625 |
| 3742.00 | 0.00513 | 0.00451 | 0.00305 | 0.00625 |
| 3747.00 | 0.00000 | 0.00909 | 0.01364 | 0.01515 |
| 1 | 4 | 16 | 4 | |
| 3746.00 | 0.00000 | 0.00909 | 0.01364 | 0.01515 |
| 3741.00 | 0.00513 | 0.00451 | 0.00305 | 0.00625 |
| 3741.00 | 0.00513 | 0.00451 | 0.00305 | 0.00625 |
| 3746.00 | 0.00000 | 0.00909 | 0.01364 | 0.01515 |
| 1 | 5 | 16 | 4 | |
| 3745.00 | 0.00000 | 0.00909 | 0.01364 | 0.01515 |
| 3740.00 | 0.00513 | 0.00451 | 0.00305 | 0.00625 |
| 3740.00 | 0.00513 | 0.00451 | 0.00305 | 0.00625 |
| 3745.00 | 0.00000 | 0.00909 | 0.01364 | 0.01515 |
| 1 | 6 | 16 | 4 | |
| 3744.00 | 0.00000 | 0.00909 | 0.01364 | 0.01515 |
| 3739.00 | 0.00513 | 0.00451 | 0.00305 | 0.00625 |
| 3739.00 | 0.00513 | 0.00451 | 0.00305 | 0.00625 |
| 3744.00 | 0.00000 | 0.00909 | 0.01364 | 0.01515 |
| 1 | 6 | 17 | 4 | |
| 3741.00 | 0.00000 | 0.00909 | 0.01364 | 0.01515 |
| 3736.00 | 0.00513 | 0.00451 | 0.00305 | 0.00625 |
| 3736.00 | 0.00513 | 0.00451 | 0.00305 | 0.00625 |
| 3741.00 | 0.00000 | 0.00909 | 0.01364 | 0.01515 |
| 1 | 6 | 18 | 4 | |
| 3737.00 | 0.00000 | 0.00909 | 0.01364 | 0.01515 |
| 3732.00 | 0.00513 | 0.00451 | 0.00305 | 0.00625 |
| 3732.00 | 0.00513 | 0.00451 | 0.00305 | 0.00625 |
| 3737.00 | 0.00000 | 0.00909 | 0.01364 | 0.01515 |
| 1 | 6 | 19 | 4 | |
| 3734.00 | 0.00000 | 0.00909 | 0.01364 | 0.01515 |
| 3729.00 | 0.00513 | 0.00451 | 0.00305 | 0.00625 |
| 3729.00 | 0.00513 | 0.00451 | 0.00305 | 0.00625 |
| 3734.00 | 0.00000 | 0.00909 | 0.01364 | 0.01515 |

```
        1        6       20        4
3730.00  0.00000  0.00909  0.01364  0.01515
3725.00  0.00513  0.00451  0.00305  0.00625
3725.00  0.00513  0.00451  0.00305  0.00625
3730.00  0.00000  0.00909  0.01364  0.01515
       25
        1        2        2        3
3796.00  0.00000  0.00000  0.00000  0.03788
3791.00  0.00000  0.00000  0.00000  0.01894
3791.00  0.00000  0.00000  0.00000  0.01894
        1        2        3        4
3796.00  0.00000  0.00000  0.00000  0.03788
3793.00  0.00000  0.00000  0.00000  0.03788
3788.00  0.00000  0.00000  0.00000  0.01894
3788.00  0.00000  0.00000  0.00000  0.01894
        1        2        4        2
3793.00  0.00000  0.00000  0.00000  0.03788
3789.00  0.00000  0.00000  0.00000  0.03788
        1        3        4        4
3784.00  0.00000  0.00000  0.00000  0.01894
3784.00  0.00000  0.00000  0.00000  0.01894
3789.00  0.00000  0.00000  0.00000  0.03788
3788.00  0.00000  0.00000  0.00000  0.03788
        1        4        4        3
3783.00  0.00000  0.00000  0.00000  0.01894
3783.00  0.00000  0.00000  0.00000  0.01894
3788.00  0.00000  0.00000  0.00000  0.03788
        1        4        5        3
3787.00  0.00000  0.00000  0.00000  0.03788
3782.00  0.00000  0.00000  0.00000  0.01894
3782.00  0.00000  0.00000  0.00000  0.01894
        1        4        6        4
3787.00  0.00000  0.00000  0.00000  0.03788
3785.00  0.00000  0.00000  0.00000  0.03788
3780.00  0.00000  0.00000  0.00000  0.01894
3780.00  0.00000  0.00000  0.00000  0.01894
        1        4        7        4
3785.00  0.00000  0.00000  0.00000  0.03788
3783.00  0.00000  0.00000  0.00000  0.03788
3778.00  0.00000  0.00000  0.00000  0.01894
3778.00  0.00000  0.00000  0.00000  0.01894
        1        4        8        3
3783.00  0.00000  0.00000  0.00000  0.03788
3778.00  0.00000  0.00000  0.00000  0.03788
3773.00  0.00000  0.00000  0.00000  0.01894
        1        4        9        3
3773.00  0.00000  0.00000  0.00000  0.01894
3778.00  0.00000  0.00000  0.00000  0.03788
3774.00  0.00000  0.00000  0.00000  0.03788
        1        4       10        3
3769.00  0.00000  0.00000  0.00000  0.01894
3769.00  0.00000  0.00000  0.00000  0.01894
3774.00  0.00000  0.00000  0.00000  0.03788
        1        4       11        4
3770.00  0.00000  0.00000  0.00000  0.03788
3765.00  0.00000  0.00000  0.00000  0.01894
3765.00  0.00000  0.00000  0.00000  0.01894
3770.00  0.00000  0.00000  0.00000  0.03788
        1        3       11        4
3767.00  0.00000  0.00000  0.00000  0.03788
3762.00  0.00000  0.00000  0.00000  0.01894
3762.00  0.00000  0.00000  0.00000  0.01894
3767.00  0.00000  0.00000  0.00000  0.03788
        1        3       12        4
3763.00  0.00000  0.00000  0.00000  0.03788
3758.00  0.00000  0.00000  0.00000  0.01894
3758.00  0.00000  0.00000  0.00000  0.01894
3763.00  0.00000  0.00000  0.00000  0.03788
```

| 1 | 3 | 13 | 4 | |
|---|---|---|---|---|
| 3762.00 | 0.00000 | 0.00000 | 0.00000 | 0.03788 |
| 3757.00 | 0.00000 | 0.00000 | 0.00000 | 0.01894 |
| 3757.00 | 0.00000 | 0.00000 | 0.00000 | 0.01894 |
| 3762.00 | 0.00000 | 0.00000 | 0.00000 | 0.03788 |
| 1 | 3 | 14 | 4 | |
| 3759.00 | 0.00000 | 0.00000 | 0.00000 | 0.03788 |
| 3754.00 | 0.00000 | 0.00000 | 0.00000 | 0.01894 |
| 3754.00 | 0.00000 | 0.00000 | 0.00000 | 0.01894 |
| 3759.00 | 0.00000 | 0.00000 | 0.00000 | 0.03788 |
| 1 | 3 | 15 | 3 | |
| 3756.00 | 0.00000 | 0.00000 | 0.00000 | 0.03788 |
| 3751.00 | 0.00000 | 0.00000 | 0.00000 | 0.01894 |
| 3751.00 | 0.00000 | 0.00000 | 0.00000 | 0.01894 |
| 1 | 3 | 16 | 3 | |
| 3756.00 | 0.00000 | 0.00000 | 0.00000 | 0.03788 |
| 3752.00 | 0.00000 | 0.00000 | 0.00000 | 0.03788 |
| 3747.00 | 0.00000 | 0.00000 | 0.00000 | 0.01894 |
| 1 | 4 | 16 | 3 | |
| 3747.00 | 0.00000 | 0.00000 | 0.00000 | 0.01894 |
| 3752.00 | 0.00000 | 0.00000 | 0.00000 | 0.03788 |
| 3748.00 | 0.00000 | 0.00000 | 0.00000 | 0.03788 |
| 1 | 5 | 16 | 3 | |
| 3743.00 | 0.00000 | 0.00000 | 0.00000 | 0.01894 |
| 3743.00 | 0.00000 | 0.00000 | 0.00000 | 0.01894 |
| 3748.00 | 0.00000 | 0.00000 | 0.00000 | 0.03788 |
| 1 | 6 | 16 | 3 | |
| 3747.00 | 0.00000 | 0.00000 | 0.00000 | 0.03788 |
| 3742.00 | 0.00000 | 0.00000 | 0.00000 | 0.01894 |
| 3742.00 | 0.00000 | 0.00000 | 0.00000 | 0.01894 |
| 1 | 6 | 17 | 3 | |
| 3747.00 | 0.00000 | 0.00000 | 0.00000 | 0.03788 |
| 3746.00 | 0.00000 | 0.00000 | 0.00000 | 0.03788 |
| 3741.00 | 0.00000 | 0.00000 | 0.00000 | 0.01894 |
| 1 | 6 | 18 | 2 | |
| 3741.00 | 0.00000 | 0.00000 | 0.00000 | 0.01894 |
| 3746.00 | 0.00000 | 0.00000 | 0.00000 | 0.03788 |
| 1 | 6 | 19 | 2 | |
| 3745.00 | 0.00000 | 0.00000 | 0.00000 | 0.03788 |
| 3740.00 | 0.00000 | 0.00000 | 0.00000 | 0.01894 |
| 1 | 6 | 20 | 2 | |
| 3740.00 | 0.00000 | 0.00000 | 0.00000 | 0.01894 |
| 3745.00 | 0.00000 | 0.00000 | 0.00000 | 0.03788 |

## 14.3 Results

Output pertinent to RIP-ET is shaded.

```
                            MODFLOW-2005
        U.S. GEOLOGICAL SURVEY MODULAR FINITE-DIFFERENCE GROUND-WATER FLOW MODEL
                        VERSION 1.8.00 12/18/2009

 LIST FILE: Exmp1.1st
                            UNIT   26

 OPENING Exmp1.bal
 FILE TYPE:BAS6   UNIT    1    STATUS:OLD
 FORMAT:FORMATTED              ACCESS:SEQUENTIAL

 OPENING Exmp1.bcf
 FILE TYPE:BCF6   UNIT   11    STATUS:OLD
 FORMAT:FORMATTED              ACCESS:SEQUENTIAL

 OPENING Exmp1.wel
 FILE TYPE:WEL    UNIT   12    STATUS:OLD
 FORMAT:FORMATTED              ACCESS:SEQUENTIAL

 OPENING Exmp1.str
 FILE TYPE:STR    UNIT   14    STATUS:OLD
 FORMAT:FORMATTED              ACCESS:SEQUENTIAL

 OPENING Exmp1.sip
 FILE TYPE:SIP    UNIT   16    STATUS:OLD
 FORMAT:FORMATTED              ACCESS:SEQUENTIAL

 OPENING Exmp1.oc
 FILE TYPE:OC    UNIT   17    STATUS:OLD
 FORMAT:FORMATTED              ACCESS:SEQUENTIAL

 OPENING Exmp1.rip
 FILE TYPE:RIP   UNIT   20    STATUS:OLD
 FORMAT:FORMATTED              ACCESS:SEQUENTIAL

 OPENING Exmp1.dis
 FILE TYPE:DIS   UNIT   21    STATUS:OLD
 FORMAT:FORMATTED   .          ACCESS:SEQUENTIAL

 OPENING Exmp1.hed
 FILE TYPE:DATA   UNIT   30    STATUS:UNKNOWN
 FORMAT:FORMATTED              ACCESS:SEQUENTIAL

 OPENING Initial.hed
 FILE TYPE:DATA   UNIT   40    STATUS:UNKNOWN
 FORMAT:FORMATTED              ACCESS:SEQUENTIAL

 OPENING LandSurface.dem
 FILE TYPE:DATA   UNIT   45    STATUS:UNKNOWN
 FORMAT:FORMATTED              ACCESS:SEQUENTIAL

 OPENING RIP_POLY_BASE.ret
 FILE TYPE:DATA   UNIT   50    STATUS:UNKNOWN
 FORMAT:FORMATTED              ACCESS:SEQUENTIAL

 BAS -- BASIC PACKAGE, VERSION 7, 5/2/2005 INPUT READ FROM UNIT    1

 DISCRETIZATION INPUT DATA READ FROM UNIT   21
 # Riparian ET Test Model, 2 Season Model
    1 LAYERS        12 ROWS         20 COLUMNS
    2 STRESS PERIOD(S) IN SIMULATION
 MODEL TIME UNIT IS SECONDS
 MODEL LENGTH UNIT IS FEET
  Confining bed flag for each layer:
   0

                    DELR =   5280.00

                    DELC =   5280.00
```

```
          TOP ELEVATION OF LAYER 1
READING ON UNIT   45 WITH FORMAT: (10F8.1)

     1      2      3      4      5      6      7      8      9     10     11     12     13     14     15     16     17     18     19     20
 ...........................................................................................................................................
  1 3800.  3799.  3789.  3787.  3784.  -999.  -999.  -999.  -999.  -999.  3760.  3756.  3756.  3749.  3744.  3743.  3740.  3739.  -999.  -999.
  2 3800.  3794.  3788.  3782.  3784.  -999.  -999.  -999.  -999.  -999.  3757.  3758.  3753.  3747.  3743.  3742.  3740.  3738.  -999.  -999.
  3 3800.  3798.  3788.  3782.  3782.  3778.  3774.  3772.  3768.  3764.  3756.  3754.  3750.  3746.  3742.  3741.  3739.  3736.  -999.  -999.
  4 -999.  -999.  3789.  3781.  3779.  3778.  3772.  3768.  3764.  3761.  3758.  3756.  3753.  3748.  3744.  3740.  3738.  3734.  -999.  -999.
  5 -999.  -999.  3790.  3782.  3782.  3780.  3772.  3769.  3767.  3763.  3760.  3759.  3753.  3749.  3746.  3739.  3737.  3733.  3728.  3724.
  6 -999.  -999.  3793.  3782.  3786.  3780.  3776.  3772.  3768.  3766.  3762.  3759.  3755.  3750.  3748.  3738.  3735.  3730.  3728.  3724.
  7 -999.  -999.  3793.  3786.  3790.  3784.  3779.  3774.  3770.  3768.  3764.  3761.  3757.  3752.  3749.  3738.  3738.  3732.  3728.  3724.
  8 -999.  -999.  3794.  3787.  3790.  3787.  3780.  3774.  3771.  3772.  3766.  3763.  3759.  3755.  3752.  3739.  3740.  3732.  3730.  3724.
  9 -999.  -999.  3796.  3790.  3794.  3790.  3783.  3775.  3774.  3773.  3769.  3764.  3760.  3756.  3754.  3740.  3743.  3734.  3730.  3725.
 10 -999.  -999.  -999.  -999.  -999.  3794.  3787.  3776.  3777.  3774.  3772.  3767.  3762.  3756.  3755.  3740.  3745.  3735.  3731.  3726.
 11 -999.  -999.  -999.  -999.  -999.  3795.  3789.  3777.  3779.  3778.  3773.  3767.  3764.  3757.  -999.  -999.  -999.  -999.  -999.  -999.
 12 -999.  -999.  -999.  -999.  -999.  3798.  3792.  3781.  3781.  3779.  3775.  3767.  3767.  3759.  -999.  -999.  -999.  -999.  -999.  -999.

  MODEL LAYER BOTTOM EL. =   3500.00      FOR LAYER   1
```

| STRESS PERIOD | LENGTH | TIME STEPS | MULTIPLIER FOR DELT | SS FLAG |
|---|---|---|---|---|
| 1 | 1.5778800E+07 | 1 | 1.000 | TR |
| 2 | 1.5778800E+07 | 1 | 1.000 | TR |

TRANSIENT SIMULATION


# Riparian ET Test Model, 2 Season Model
THE FREE FORMAT OPTION HAS BEEN SELECTED

```
                 BOUNDARY ARRAY FOR LAYER   1
READING ON UNIT   1 WITH FORMAT: (20I3)

    1  2  3  4  5  6  7  8  9 10 11 12 13 14 15 16 17 18 19 20
 ...........................................................
  1 -1  1  1  1  1  1  0  0  0  0  0  1  1  1  1  1  1  1  0  0
  2 -1  1  1  1  1  0  0  0  0  0  1  1  1  1  1  1  1  1  0  0
  3 -1  1  1  1  1  1  1  1  1  1  1  1  1  1  1  1  1  1  0  0
  4  0  0  1  1  1  1  1  1  1  1  1  1  1  1  1  1  1  1  0  0
  5  0  0  1  1  1  1  1  1  1  1  1  1  1  1  1  1  1  1  1  1
  6  0  0  1  1  1  1  1  1  1  1  1  1  1  1  1  1  1  1  1  1
  7  0  0  1  1  1  1  1  1  1  1  1  1  1  1  1  1  1  1  1  1
  8  0  0  1  1  1  1  1  1  1  1  1  1  1  1  1  1  1  1  1  1
  9  0  0  1  1  1  1  1  1  1  1  1  1  1  1  1  1  1  1  1  1
 10  0  0  0  0  0  1  1  1  1  1  1  1  1  1  1  1  1  1  1  1
 11  0  0  0  0  0  1  1  1  1  1  1  1  1  1  0  0  0  0  0  0
 12  0  0  0  0  0  1  1  1  1  1  1  1  1  1  0  0  0  0  0  0

 AQUIFER HEAD WILL BE SET TO  -999.00    AT ALL NO-FLOW NODES (IBOUND=0).
```

```
                      INITIAL HEAD FOR LAYER   1
READING ON UNIT   40 WITH FORMAT: (10F8.1)

        1       2       3       4       5       6       7       8       9      10      11      12      13      14      15
       16      17      18      19      20

 1   3800.0  3793.5  3788.0  3784.0  3781.9  -999.0  -999.0  -999.0  -999.0  -999.0  3746.3  3745.3  3743.5  3741.5  3739.5
     3737.7  3736.3  3735.6  -999.0  -999.0
 2   3800.0  3792.5  3786.4  3782.1  3779.8  -999.0  -999.0  -999.0  -999.0  -999.0  3747.4  3745.9  3743.8  3741.5  3739.3
     3737.4  3735.7  3734.8  -999.0  -999.0
 3   3800.0  3792.2  3783.8  3779.2  3775.4  3769.9  3765.6  3761.6  3757.8  3754.1  3750.1  3747.1  3744.3  3741.5  3738.9
     3736.7  3734.5  3733.1  -999.0  -999.0
 4   -999.0  -999.0  3777.2  3775.3  3772.5  3768.9  3765.1  3761.4  3757.8  3754.3  3750.7  3747.0  3743.9  3740.9  3738.0
     3736.4  3732.4  3730.0  -999.0  -999.0
 5   -999.0  -999.0  3772.6  3771.3  3769.1  3766.3  3763.1  3759.9  3756.6  3753.3  3749.9  3746.5  3743.3  3740.1  3736.9
     3733.7  3729.6  3724.5  3715.8  3707.0
 6   -999.0  -999.0  3769.1  3768.1  3766.4  3764.0  3761.3  3758.4  3755.4  3752.3  3749.2  3745.9  3742.6  3739.2  3735.7
     3732.1  3727.7  3722.3  3715.4  3707.3
 7   -999.0  -999.0  3766.6  3765.8  3764.2  3762.0  3759.6  3757.0  3754.3  3751.4  3748.4  3745.3  3742.0  3738.4  3734.5
     3730.2  3725.2  3719.3  3712.2  3703.9
 8   -999.0  -999.0  3764.9  3764.1  3762.5  3760.3  3758.1  3755.8  3753.3  3750.7  3747.9  3744.8  3741.6  3737.9  3733.6
     3728.8  3723.4  3717.1  3709.8  3701.4
 9   -999.0  -999.0  3764.1  3763.2  3761.5  3758.7  3756.6  3754.6  3752.4  3750.0  3747.4  3744.6  3741.5  3737.8  3733.2
     3728.0  3722.3  3715.8  3708.3  3699.8
10   -999.0  -999.0  -999.0  -999.0  -999.0  3756.2  3755.1  3753.6  3751.7  3749.6  3747.2  3744.7  3742.0  3738.6  3733.2
     3727.7  3721.8  3715.1  3707.6  3699.0
11   -999.0  -999.0  -999.0  -999.0  -999.0  3754.8  3754.1  3752.8  3751.2  3749.3  3747.2  3745.1  3743.0  3741.4  -999.0
     -999.0  -999.0  -999.0  -999.0  -999.0
12   -999.0  -999.0  -999.0  -999.0  -999.0  3754.2  3753.6  3752.4  3750.9  3749.1  3747.2  3745.3  3743.6  3742.5  -999.0
     -999.0  -999.0  -999.0  -999.0  -999.0
```

OUTPUT CONTROL IS SPECIFIED ONLY AT TIME STEPS FOR WHICH OUTPUT IS DESIRED
HEADS WILL BE SAVED WITH FORMAT: (10F8.1)
COMPACT CELL-BY-CELL BUDGET FILES WILL BE WRITTEN
HEAD PRINT FORMAT CODE IS   7   DRAWDOWN PRINT FORMAT CODE IS   0
HEADS WILL BE SAVED ON UNIT   30   DRAWDOWNS WILL BE SAVED ON UNIT    0

BCF -- BLOCK-CENTERED FLOW PACKAGE, VERSION 7, 5/2/2005
        INPUT READ FROM UNIT 11
TRANSIENT SIMULATION
CONSTANT-HEAD CELL-BY-CELL FLOWS WILL BE PRINTED WHEN ICBCFL IS NOT 0
HEAD AT CELLS THAT CONVERT TO DRY=  0.10000E+31
WETTING CAPABILITY IS NOT ACTIVE
      LAYER  LAYER-TYPE CODE     INTERBLOCK T
      ----------------------------------------------
        1            1         0 -- HARMONIC

COLUMN TO ROW ANISOTROPY =   1.00000

   PRIMARY STORAGE COEF =  1.000000E-02 FOR LAYER    1

   HYD. COND. ALONG ROWS =  1.060000E-03 FOR LAYER    1

WEL -- WELL PACKAGE, VERSION 7, 5/2/2005 INPUT READ FROM UNIT   12
No named parameters
MAXIMUM OF      6 ACTIVE WELLS AT ONE TIME
CELL-BY-CELL FLOWS WILL BE PRINTED WHEN ICBCFL NOT 0


     0 Well parameters

```
RIP-ET -- RIPARIAN PACKAGE, VERSION 3, 6/21/2010 INPUT READ FROM UNIT 20
MAXIMUM OF   25 RIPARIAN CELLS
MAXIMUM OF    4 POLYGONS PER CELL
THE ET RATE FOR EACH PLANT FUNCTIONAL SUBGROUP WILL BE WRITTEN TO THE LIST FILE WHEN ICBCFL IS NOT 0
CELL LOCATION, SURFACE ELEVATION, AND PLANT FUNCTIONAL SUBGROUP ET RATES WILL BE SAVED ON UNIT 50
MAXIMUM NUMBERS OF PLANT FUNCTIONAL SUBGROUPS AND SEGMENTS
PLANT FUNCTIONAL SUBGROUPS  = 4
MAXIMUM CURVE SEGMENTS      = 7
```

                        RIPARIAN INFORMATION

| NAME | SATURATION EXTINCTION DEPTH | ACTIVE ROOT DEPTH | MAXIMUM ET FLUX | ET FLUX AT SATURATION EXTINCTION DEPTH |
|------|------|------|------|------|
| D.R. Riparian Small | 0.0000 | 13.1200 | 0.7610E-07 | 0.0000E+00 |
| D.R. Riparian Medium | 0.0000 | 16.4000 | 0.9910E-07 | 0.0000E+00 |
| D.R. Riparian Large | 0.0000 | 16.4000 | 0.1120E-06 | 0.0000E+00 |
| Evaporation | 0.0000 | 3.2800 | 0.2090E-06 | 0.2090E-06 |

                    SEGMENT INFORMATION

| NAME | | SEGMENTS | | | | | |
|------|------|------|------|------|------|------|------|
| D.R. Riparian Small | fdh | 0.2500 | 0.2500 | 0.1250 | 0.1250 | 0.0625 | 0.1875 |
| | dfR | 0.2501 | 0.2496 | 0.3000 | 0.2002 | -0.1303 | -0.8697 |
| D.R. Riparian Medium | fdh | 0.2000 | 0.2000 | 0.2000 | 0.1000 | 0.1000 | 0.0500 | 0.1500 |
| | dfR | 0.3218 | 0.3103 | 0.3065 | 0.0613 | -0.0958 | -0.3257 | -0.5785 |
| D.R. Riparian Large | fdh | 0.2000 | 0.2000 | 0.2000 | 0.1000 | 0.1000 | 0.0500 | 0.1500 |
| | dfR | 0.3605 | 0.3605 | 0.2361 | 0.0628 | -0.0988 | -0.4256 | -0.4755 |
| Evaporation | fdh | 1.0000 | | | | | |
| | dfR | 1.0000 | | | | | |

```
STR -- STREAM PACKAGE, VERSION 7, 5/2/2005 INPUT READ FROM UNIT   14
No named parameters
MAXIMUM OF    25 ACTIVE STREAM NODES AT ONE TIME
NUMBER OF STREAM SEGMENTS IS      1
NUMBER OF STREAM TRIBUTARIES IS        0
STREAM STAGES WILL BE CALCULATED USING A CONSTANT OF     1.4860


    0 Stream parameters

SIP -- STRONGLY-IMPLICIT PROCEDURE SOLUTION PACKAGE
            VERSION 7, 5/2/2005 INPUT READ FROM UNIT   16
MAXIMUM OF 100 ITERATIONS ALLOWED FOR CLOSURE
 5 ITERATION PARAMETERS

        SOLUTION BY THE STRONGLY IMPLICIT PROCEDURE
        ----------------------------------------------
MAXIMUM ITERATIONS ALLOWED FOR CLOSURE =     100
            ACCELERATION PARAMETER =      1.0000
    HEAD CHANGE CRITERION FOR CLOSURE =    0.10000E-03
    SIP HEAD CHANGE PRINTOUT INTERVAL =     999

    CALCULATE ITERATION PARAMETERS FROM MODEL CALCULATED WSEED
1
                        STRESS PERIOD NO.   1, LENGTH =  0.1577880E+08
                        ----------------------------------------------

                    NUMBER OF TIME STEPS =     1

                    MULTIPLIER FOR DELT =    1.000

                    INITIAL TIME STEP SIZE =  0.1577880E+08

WELL NO.  LAYER   ROW   COL   STRESS RATE
----------------------------------------------
        1      1     5    20    -1.000
        2      1     6    20    -1.000
        3      1     7    20    -1.000
        4      1     8    20    -1.000
        5      1     9    20    -1.000
        6      1    10    20    -1.000

    6 WELLS
```

25 RIPARIAN CELLS.

RIPARIAN CELL INFORMATION

| LAYER | ROW | COLUMN | POLYGON NUMBER | SURFACE ELEVATION | FRACTIONAL COVERAGE OF PLANT FUNCTIONAL SUBGROUPS D.R. Riparian Small | D.R. Riparian Medium | D.R. Riparian Large | Evaporation |
|---|---|---|---|---|---|---|---|---|
| 1 | 2 | 2 | 1 | 3796.00 | 0.00000 | 0.00909 | 0.01364 | 0.01515 |
| | | | 2 | 3791.00 | 0.00513 | 0.00451 | 0.00305 | 0.00625 |
| | | | 3 | 3791.00 | 0.00513 | 0.00451 | 0.00305 | 0.00625 |
| | | | 4 | 3796.00 | 0.00000 | 0.00909 | 0.01364 | 0.01515 |
| 1 | 2 | 3 | 1 | 3793.00 | 0.00000 | 0.00909 | 0.01364 | 0.01515 |
| | | | 2 | 3788.00 | 0.00513 | 0.00451 | 0.00305 | 0.00625 |
| | | | 3 | 3788.00 | 0.00513 | 0.00451 | 0.00305 | 0.00625 |
| | | | 4 | 3793.00 | 0.00000 | 0.00909 | 0.01364 | 0.01515 |
| 1 | 2 | 4 | 1 | 3789.00 | 0.00000 | 0.00909 | 0.01364 | 0.01515 |
| | | | 2 | 3784.00 | 0.00513 | 0.00451 | 0.00305 | 0.00625 |
| | | | 3 | 3784.00 | 0.00513 | 0.00451 | 0.00305 | 0.00625 |
| | | | 4 | 3789.00 | 0.00000 | 0.00909 | 0.01364 | 0.01515 |
| 1 | 3 | 4 | 1 | 3788.00 | 0.00000 | 0.00909 | 0.01364 | 0.01515 |
| | | | 2 | 3783.00 | 0.00513 | 0.00451 | 0.00305 | 0.00625 |
| | | | 3 | 3783.00 | 0.00513 | 0.00451 | 0.00305 | 0.00625 |
| | | | 4 | 3788.00 | 0.00000 | 0.00909 | 0.01364 | 0.01515 |
| 1 | 4 | 4 | 1 | 3787.00 | 0.00000 | 0.00909 | 0.01364 | 0.01515 |
| | | | 2 | 3782.00 | 0.00513 | 0.00451 | 0.00305 | 0.00625 |
| | | | 3 | 3782.00 | 0.00513 | 0.00451 | 0.00305 | 0.00625 |
| | | | 4 | 3787.00 | 0.00000 | 0.00909 | 0.01364 | 0.01515 |
| 1 | 4 | 5 | 1 | 3785.00 | 0.00000 | 0.00909 | 0.01364 | 0.01515 |
| | | | 2 | 3780.00 | 0.00513 | 0.00451 | 0.00305 | 0.00625 |
| | | | 3 | 3780.00 | 0.00513 | 0.00451 | 0.00305 | 0.00625 |
| | | | 4 | 3785.00 | 0.00000 | 0.00909 | 0.01364 | 0.01515 |
| 1 | 4 | 6 | 1 | 3783.00 | 0.00000 | 0.00909 | 0.01364 | 0.01515 |
| | | | 2 | 3778.00 | 0.00513 | 0.00451 | 0.00305 | 0.00625 |
| | | | 3 | 3778.00 | 0.00513 | 0.00451 | 0.00305 | 0.00625 |
| | | | 4 | 3783.00 | 0.00000 | 0.00909 | 0.01364 | 0.01515 |
| 1 | 4 | 7 | 1 | 3778.00 | 0.00000 | 0.00909 | 0.01364 | 0.01515 |
| | | | 2 | 3773.00 | 0.00513 | 0.00451 | 0.00305 | 0.00625 |
| | | | 3 | 3773.00 | 0.00513 | 0.00451 | 0.00305 | 0.00625 |
| | | | 4 | 3778.00 | 0.00000 | 0.00909 | 0.01364 | 0.01515 |
| 1 | 4 | 8 | 1 | 3774.00 | 0.00000 | 0.00909 | 0.01364 | 0.01515 |
| | | | 2 | 3769.00 | 0.00513 | 0.00451 | 0.00305 | 0.00625 |
| | | | 3 | 3769.00 | 0.00513 | 0.00451 | 0.00305 | 0.00625 |
| | | | 4 | 3774.00 | 0.00000 | 0.00909 | 0.01364 | 0.01515 |
| 1 | 4 | 9 | 1 | 3770.00 | 0.00000 | 0.00909 | 0.01364 | 0.01515 |
| | | | 2 | 3765.00 | 0.00513 | 0.00451 | 0.00305 | 0.00625 |
| | | | 3 | 3765.00 | 0.00513 | 0.00451 | 0.00305 | 0.00625 |
| | | | 4 | 3770.00 | 0.00000 | 0.00909 | 0.01364 | 0.01515 |
| 1 | 4 | 10 | 1 | 3767.00 | 0.00000 | 0.00909 | 0.01364 | 0.01515 |
| | | | 2 | 3762.00 | 0.00513 | 0.00451 | 0.00305 | 0.00625 |
| | | | 3 | 3762.00 | 0.00513 | 0.00451 | 0.00305 | 0.00625 |
| | | | 4 | 3767.00 | 0.00000 | 0.00909 | 0.01364 | 0.01515 |
| 1 | 4 | 11 | 1 | 3763.00 | 0.00000 | 0.00909 | 0.01364 | 0.01515 |
| | | | 2 | 3758.00 | 0.00513 | 0.00451 | 0.00305 | 0.00625 |
| | | | 3 | 3758.00 | 0.00513 | 0.00451 | 0.00305 | 0.00625 |
| | | | 4 | 3763.00 | 0.00000 | 0.00909 | 0.01364 | 0.01515 |
| 1 | 3 | 11 | 1 | 3762.00 | 0.00000 | 0.00909 | 0.01364 | 0.01515 |
| | | | 2 | 3757.00 | 0.00513 | 0.00451 | 0.00305 | 0.00625 |
| | | | 3 | 3757.00 | 0.00513 | 0.00451 | 0.00305 | 0.00625 |
| | | | 4 | 3762.00 | 0.00000 | 0.00909 | 0.01364 | 0.01515 |
| 1 | 3 | 12 | 1 | 3759.00 | 0.00000 | 0.00909 | 0.01364 | 0.01515 |
| | | | 2 | 3754.00 | 0.00513 | 0.00451 | 0.00305 | 0.00625 |
| | | | 3 | 3754.00 | 0.00513 | 0.00451 | 0.00305 | 0.00625 |
| | | | 4 | 3759.00 | 0.00000 | 0.00909 | 0.01364 | 0.01515 |
| 1 | 3 | 13 | 1 | 3756.00 | 0.00000 | 0.00909 | 0.01364 | 0.01515 |
| | | | 2 | 3751.00 | 0.00513 | 0.00451 | 0.00305 | 0.00625 |
| | | | 3 | 3751.00 | 0.00513 | 0.00451 | 0.00305 | 0.00625 |
| | | | 4 | 3756.00 | 0.00000 | 0.00909 | 0.01364 | 0.01515 |
| 1 | 3 | 14 | 1 | 3752.00 | 0.00000 | 0.00909 | 0.01364 | 0.01515 |
| | | | 2 | 3747.00 | 0.00513 | 0.00451 | 0.00305 | 0.00625 |
| | | | 3 | 3747.00 | 0.00513 | 0.00451 | 0.00305 | 0.00625 |
| | | | 4 | 3752.00 | 0.00000 | 0.00909 | 0.01364 | 0.01515 |
| 1 | 3 | 15 | 1 | 3748.00 | 0.00000 | 0.00909 | 0.01364 | 0.01515 |
| | | | 2 | 3743.00 | 0.00513 | 0.00451 | 0.00305 | 0.00625 |
| | | | 3 | 3743.00 | 0.00513 | 0.00451 | 0.00305 | 0.00625 |
| | | | 4 | 3748.00 | 0.00000 | 0.00909 | 0.01364 | 0.01515 |
| 1 | 3 | 16 | 1 | 3747.00 | 0.00000 | 0.00909 | 0.01364 | 0.01515 |
| | | | 2 | 3742.00 | 0.00513 | 0.00451 | 0.00305 | 0.00625 |
| | | | 3 | 3742.00 | 0.00513 | 0.00451 | 0.00305 | 0.00625 |
| | | | 4 | 3747.00 | 0.00000 | 0.00909 | 0.01364 | 0.01515 |
| 1 | 4 | 16 | 1 | 3746.00 | 0.00000 | 0.00909 | 0.01364 | 0.01515 |
| | | | 2 | 3741.00 | 0.00513 | 0.00451 | 0.00305 | 0.00625 |
| | | | 3 | 3741.00 | 0.00513 | 0.00451 | 0.00305 | 0.00625 |
| | | | 4 | 3746.00 | 0.00000 | 0.00909 | 0.01364 | 0.01515 |
| 1 | 5 | 16 | 1 | 3745.00 | 0.00000 | 0.00909 | 0.01364 | 0.01515 |
| | | | 2 | 3740.00 | 0.00513 | 0.00451 | 0.00305 | 0.00625 |
| | | | 3 | 3740.00 | 0.00513 | 0.00451 | 0.00305 | 0.00625 |
| | | | 4 | 3745.00 | 0.00000 | 0.00909 | 0.01364 | 0.01515 |

| LAYER | ROW | COL | | | | | | | |
|---|---|---|---|---|---|---|---|---|---|
| 1 | 6 | 16 | 1 | 3744.00 | 0.00000 | | 0.00909 | 0.01364 | 0.01515 |
| | | | 2 | 3739.00 | 0.00513 | | 0.00451 | 0.00306 | 0.00625 |
| | | | 3 | 3739.00 | 0.00513 | | 0.00451 | 0.00305 | 0.00625 |
| | | | 4 | 3744.00 | 0.00000 | | 0.00909 | 0.01364 | 0.01515 |
| 1 | 6 | 17 | 1 | 3741.00 | 0.00000 | | 0.00909 | 0.01364 | 0.01515 |
| | | | 2 | 3736.00 | 0.00513 | | 0.00451 | 0.00305 | 0.00625 |
| | | | 3 | 3736.00 | 0.00513 | | 0.00451 | 0.00305 | 0.00625 |
| | | | 4 | 3741.00 | 0.00000 | | 0.00909 | 0.01364 | 0.01515 |
| 1 | 6 | 18 | 1 | 3737.00 | 0.00000 | | 0.00909 | 0.01364 | 0.01515 |
| | | | 2 | 3732.00 | 0.00513 | | 0.00451 | 0.00305 | 0.00625 |
| | | | 3 | 3732.00 | 0.00513 | | 0.00451 | 0.00305 | 0.00625 |
| | | | 4 | 3737.00 | 0.00000 | | 0.00909 | 0.01364 | 0.01515 |
| 1 | 6 | 19 | 1 | 3734.00 | 0.00000 | | 0.00909 | 0.01364 | 0.01515 |
| | | | 2 | 3729.00 | 0.00513 | | 0.00451 | 0.00306 | 0.00625 |
| | | | 3 | 3729.00 | 0.00513 | | 0.00451 | 0.00305 | 0.00625 |
| | | | 4 | 3734.00 | 0.00000 | | 0.00909 | 0.01364 | 0.01515 |
| 1 | 6 | 20 | 1 | 3730.00 | 0.00000 | | 0.00909 | 0.01364 | 0.01515 |
| | | | 2 | 3725.00 | 0.00513 | | 0.00451 | 0.00305 | 0.00625 |
| | | | 3 | 3725.00 | 0.00513 | | 0.00451 | 0.00305 | 0.00625 |
| | | | 4 | 3730.00 | 0.00000 | | 0.00909 | 0.01364 | 0.01515 |

| LAYER | ROW | COL | SEGMENT NUMBER | REACH NUMBER | STREAMFLOW | STREAM STAGE | STREAMBED CONDUCTANCE | STREAMBED BOT ELEVATION | STREAMBED TOP ELEVATION |
|---|---|---|---|---|---|---|---|---|---|
| 1 | 2 | 2 | 1 | 1 | 100.0 | 3796. | 0.9240E-01 | 3776. | 3786. |
| 1 | 2 | 3 | 1 | 2 | 0.000 | 0.000 | 0.9240E-01 | 3773. | 3783. |
| 1 | 2 | 4 | 1 | 3 | 0.000 | 0.000 | 0.9240E-01 | 3769. | 3779. |
| 1 | 3 | 4 | 1 | 4 | 0.000 | 0.000 | 0.9240E-01 | 3768. | 3778. |
| 1 | 4 | 4 | 1 | 5 | 0.000 | 0.000 | 0.9240E-01 | 3767. | 3777. |
| 1 | 4 | 5 | 1 | 6 | 0.000 | 0.000 | 0.9240E-01 | 3765. | 3775. |
| 1 | 4 | 6 | 1 | 7 | 0.000 | 0.000 | 0.9240E-01 | 3763. | 3773. |
| 1 | 4 | 7 | 1 | 8 | 0.000 | 0.000 | 0.9240E-01 | 3758. | 3768 |
| 1 | 4 | 8 | 1 | 9 | 0.000 | 0.000 | 0.9240E-01 | 3754. | 3764. |
| 1 | 4 | 9 | 1 | 10 | 0.000 | 0.000 | 0.9240E-01 | 3760. | 3760. |
| 1 | 4 | 10 | 1 | 11 | 0.000 | 0.000 | 0.9240E-01 | 3747. | 3757. |
| 1 | 4 | 11 | 1 | 12 | 0.000 | 0.000 | 0.9240E-01 | 3743. | 3753. |
| 1 | 3 | 11 | 1 | 13 | 0.000 | 0.000 | 0.9240E-01 | 3742. | 3752. |
| 1 | 3 | 12 | 1 | 14 | 0.000 | 0.000 | 0.9240E-01 | 3739. | 3749. |
| 1 | 3 | 13 | 1 | 15 | 0.000 | 0.000 | 0.9240E-01 | 3736. | 3746. |
| 1 | 3 | 14 | 1 | 16 | 0.000 | 0.000 | 0.9240E-01 | 3732. | 3742. |
| 1 | 3 | 15 | 1 | 17 | 0.000 | 0.000 | 0.9240E-01 | 3728. | 3738. |
| 1 | 3 | 16 | 1 | 18 | 0.000 | 0.000 | 0.9240E-01 | 3727. | 3737. |
| 1 | 4 | 16 | 1 | 19 | 0.000 | 0.000 | 0.9240E-01 | 3726. | 3736. |
| 1 | 5 | 16 | 1 | 20 | 0.000 | 0.000 | 0.9240E-01 | 3725. | 3735. |
| 1 | 6 | 16 | 1 | 21 | 0.000 | 0.000 | 0.9240E-01 | 3724. | 3734. |
| 1 | 6 | 17 | 1 | 22 | 0.000 | 0.000 | 0.9240E-01 | 3721. | 3731. |
| 1 | 6 | 18 | 1 | 23 | 0.000 | 0.000 | 0.9240E-01 | 3717. | 3727. |
| 1 | 6 | 19 | 1 | 24 | 0.000 | 0.000 | 0.9240E-01 | 3714. | 3724. |
| 1 | 6 | 20 | 1 | 25 | 0.000 | 0.000 | 0.9240E-01 | 3710. | 3720. |

25 STREAM REACHES

| LAYER | ROW | COL | SEGMENT NUMBER | REACH NUMBER | STREAM WIDTH | STREAM SLOPE | ROUGH COEF. |
|---|---|---|---|---|---|---|---|
| 1 | 2 | 2 | 1 | 1 | 100.0 | 0.5680E-03 | 0.3000E-01 |
| 1 | 2 | 3 | 1 | 2 | 100.0 | 0.7580E-03 | 0.3000E-01 |
| 1 | 2 | 4 | 1 | 3 | 100.0 | 0.1890E-03 | 0.3000E-01 |
| 1 | 3 | 4 | 1 | 4 | 100.0 | 0.1890E-03 | 0.3000E-01 |
| 1 | 4 | 4 | 1 | 5 | 100.0 | 0.3790E-03 | 0.3000E-01 |
| 1 | 4 | 5 | 1 | 6 | 100.0 | 0.3790E-03 | 0.3000E-01 |
| 1 | 4 | 6 | 1 | 7 | 100.0 | 0.9470E-03 | 0.3000E-01 |
| 1 | 4 | 7 | 1 | 8 | 100.0 | 0.7580E-03 | 0.3000E-01 |
| 1 | 4 | 8 | 1 | 9 | 100.0 | 0.7580E-03 | 0.3000E-01 |
| 1 | 4 | 9 | 1 | 10 | 100.0 | 0.5680E-03 | 0.3000E-01 |
| 1 | 4 | 10 | 1 | 11 | 100.0 | 0.7580E-03 | 0.3000E-01 |
| 1 | 4 | 11 | 1 | 12 | 100.0 | 0.1890E-03 | 0.3000E-01 |
| 1 | 3 | 11 | 1 | 13 | 100.0 | 0.1890E-03 | 0.3000E-01 |
| 1 | 3 | 12 | 1 | 14 | 100.0 | 0.5680E-03 | 0.3000E-01 |
| 1 | 3 | 13 | 1 | 15 | 100.0 | 0.5680E-03 | 0.3000E-01 |
| 1 | 3 | 14 | 1 | 16 | 100.0 | 0.7580E-03 | 0.3000E-01 |
| 1 | 3 | 15 | 1 | 17 | 100.0 | 0.7580E-03 | 0.3000E-01 |
| 1 | 3 | 16 | 1 | 18 | 100.0 | 0.1890E-03 | 0.3000E-01 |
| 1 | 4 | 16 | 1 | 19 | 100.0 | 0.1890E-03 | 0.3000E-01 |
| 1 | 5 | 16 | 1 | 20 | 100.0 | 0.1890E-03 | 0.3000E-01 |
| 1 | 6 | 16 | 1 | 21 | 100.0 | 0.1890E-03 | 0.3000E-01 |
| 1 | 6 | 17 | 1 | 22 | 100.0 | 0.5680E-03 | 0.3000E-01 |
| 1 | 6 | 18 | 1 | 23 | 100.0 | 0.7580E-03 | 0.3000E-01 |
| 1 | 6 | 19 | 1 | 24 | 100.0 | 0.5680E-03 | 0.3000E-01 |
| 1 | 6 | 20 | 1 | 25 | 100.0 | 0.7580E-03 | 0.3000E-01 |

SOLVING FOR HEAD

AVERAGE SEED = 0.00614438
MINIMUM SEED = 0.00606986

```
    5 ITERATION PARAMETERS CALCULATED FROM AVERAGE SEED:

 0.000000E+00 0.720025E+00 0.921614E+00 0.978054E+00 0.993856E+00

    14 ITERATIONS FOR TIME STEP   1 IN STRESS PERIOD    1

MAXIMUM HEAD CHANGE FOR EACH ITERATION:

   HEAD CHANGE     HEAD CHANGE     HEAD CHANGE     HEAD CHANGE     HEAD CHANGE
  LAYER,ROW,COL   LAYER,ROW,COL   LAYER,ROW,COL   LAYER,ROW,COL   LAYER,ROW,COL
 ----------------------------------------------------------------------------
    6.424           2.887           2.325          0.5276          0.9905E-01
 (  1, 10, 20)   (  1,  9, 19)   (  1, 10, 20)   (  1, 10, 20)   (  1,  8, 12)
    0.1699E-01      0.1010E-01      0.1153E-01      0.4956E-02      0.2551E-02
 (  1, 10, 15)   (  1, 12, 13)   (  1, 11, 14)   (  1, 12, 11)   (  1, 11, 14)
   -0.2892E-03      0.1439E-03      0.1407E-03      0.7265E-04
 (  1, 10, 17)   (  1, 11, 13)   (  1, 12, 13)   (  1, 11, 14)


OUTPUT CONTROL FOR STRESS PERIOD    1   TIME STEP   1
    PRINT HEAD FOR ALL LAYERS
    PRINT BUDGET
    SAVE BUDGET
    SAVE HEAD FOR ALL LAYERS

    CONSTANT HEAD    PERIOD  1    STEP  1
 LAYER  1   ROW   1   COL   1   RATE   2.04903
 LAYER  1   ROW   2   COL   1   RATE   2.37728
 LAYER  1   ROW   3   COL   1   RATE   2.47383

         WELLS    PERIOD   1    STEP   1
 WELL    1   LAYER  1   ROW   5   COL  20   RATE  -1.00000
 WELL    2   LAYER  1   ROW   6   COL  20   RATE  -1.00000
 WELL    3   LAYER  1   ROW   7   COL  20   RATE  -1.00000
 WELL    4   LAYER  1   ROW   8   COL  20   RATE  -1.00000
 WELL    5   LAYER  1   ROW   9   COL  20   RATE  -1.00000
 WELL    6   LAYER  1   ROW  10   COL  20   RATE  -1.00000
```

```
RIPARIAN ET   PERIOD  1   STEP  1
```

| RIPARIAN ET LAYER | ROW | COLUMN | HEAD | CELL ET RATE | POLYGON NUMBER | SURFACE ELEVATION | CELL ET RATE BY PLANT FUNCTIONAL GROUP | | | |
|---|---|---|---|---|---|---|---|---|---|---|
| | | | | | | | D.R. Riparian Small | D.R. Riparian Medium | D.R. Riparian Large | Evaporation |
| 1 | 2 | 2 | 3792.35 | -0.19796 | | | | | | |
| | | | | | 1 | 3796.00 | 0.00000 | -0.02324 | -0.03932 | 0.00000 |
| | | | | | 2 | 3791.00 | 0.00000 | 0.00000 | 0.00000 | -0.03642 |
| | | | | | 3 | 3791.00 | 0.00000 | 0.00000 | 0.00000 | -0.03642 |
| | | | | | 4 | 3796.00 | 0.00000 | -0.02324 | -0.03932 | 0.00000 |
| 1 | 2 | 3 | 3786.06 | -0.18610 | | | | | | |
| | | | | | 1 | 3793.00 | 0.00000 | -0.02269 | -0.03876 | 0.00000 |
| | | | | | 2 | 3788.00 | -0.00745 | -0.00568 | -0.00357 | -0.01491 |
| | | | | | 3 | 3788.00 | -0.00745 | -0.00568 | -0.00357 | -0.01491 |
| | | | | | 4 | 3793.00 | 0.00000 | -0.02269 | -0.03876 | 0.00000 |
| 1 | 2 | 4 | 3781.77 | -0.18141 | | | | | | |
| | | | | | 1 | 3789.00 | 0.00000 | -0.02199 | -0.03785 | 0.00000 |
| | | | | | 2 | 3784.00 | -0.00860 | -0.00655 | -0.00411 | -0.01161 |
| | | | | | 3 | 3784.00 | -0.00860 | -0.00655 | -0.00411 | -0.01161 |
| | | | | | 4 | 3789.00 | 0.00000 | -0.02199 | -0.03785 | 0.00000 |
| 1 | 3 | 4 | 3778.74 | -0.15920 | | | | | | |
| | | | | | 1 | 3788.00 | 0.00000 | -0.01725 | -0.03165 | 0.00000 |
| | | | | | 2 | 3783.00 | -0.00959 | -0.01198 | -0.00914 | 0.00000 |
| | | | | | 3 | 3783.00 | -0.00959 | -0.01198 | -0.00914 | 0.00000 |
| | | | | | 4 | 3788.00 | 0.00000 | -0.01725 | -0.03165 | 0.00000 |
| 1 | 4 | 4 | 3774.90 | -0.10929 | | | | | | |
| | | | | | 1 | 3787.00 | 0.00000 | -0.01049 | -0.01955 | 0.00000 |
| | | | | | 2 | 3782.00 | -0.00499 | -0.01106 | -0.00855 | 0.00000 |
| | | | | | 3 | 3782.00 | -0.00499 | -0.01106 | -0.00855 | 0.00000 |
| | | | | | 4 | 3787.00 | 0.00000 | -0.01049 | -0.01955 | 0.00000 |
| 1 | 4 | 5 | 3772.10 | -0.09393 | | | | | | |
| | | | | | 1 | 3785.00 | 0.00000 | -0.00860 | -0.01591 | 0.00000 |
| | | | | | 2 | 3780.00 | -0.00433 | -0.01013 | -0.00800 | 0.00000 |
| | | | | | 3 | 3780.00 | -0.00433 | -0.01013 | -0.00800 | 0.00000 |
| | | | | | 4 | 3785.00 | 0.00000 | -0.00860 | -0.01591 | 0.00000 |
| 1 | 4 | 6 | 3768.53 | -0.06368 | | | | | | |
| | | | | | 1 | 3783.00 | 0.00000 | -0.00476 | -0.00880 | 0.00000 |
| | | | | | 2 | 3778.00 | -0.00303 | -0.00831 | -0.00693 | 0.00000 |
| | | | | | 3 | 3778.00 | -0.00303 | -0.00831 | -0.00693 | 0.00000 |
| | | | | | 4 | 3783.00 | 0.00000 | -0.00476 | -0.00880 | 0.00000 |
| 1 | 4 | 7 | 3764.73 | -0.08696 | | | | | | |
| | | | | | 1 | 3778.00 | 0.00000 | -0.00772 | -0.01427 | 0.00000 |
| | | | | | 2 | 3773.00 | -0.00403 | -0.00971 | -0.00776 | 0.00000 |
| | | | | | 3 | 3773.00 | -0.00403 | -0.00971 | -0.00776 | 0.00000 |
| | | | | | 4 | 3778.00 | 0.00000 | -0.00772 | -0.01427 | 0.00000 |
| 1 | 4 | 8 | 3761.06 | -0.09331 | | | | | | |
| | | | | | 1 | 3774.00 | 0.00000 | -0.00852 | -0.01576 | 0.00000 |
| | | | | | 2 | 3769.00 | -0.00430 | -0.01009 | -0.00798 | 0.00000 |
| | | | | | 3 | 3769.00 | -0.00430 | -0.01009 | -0.00798 | 0.00000 |
| | | | | | 4 | 3774.00 | 0.00000 | -0.00852 | -0.01576 | 0.00000 |
| 1 | 4 | 9 | 3757.48 | -0.10130 | | | | | | |
| | | | | | 1 | 3770.00 | 0.00000 | -0.00951 | -0.01766 | 0.00000 |
| | | | | | 2 | 3765.00 | -0.00464 | -0.01058 | -0.00827 | 0.00000 |
| | | | | | 3 | 3765.00 | -0.00464 | -0.01058 | -0.00827 | 0.00000 |
| | | | | | 4 | 3770.00 | 0.00000 | -0.00951 | -0.01766 | 0.00000 |

| | | | | | | | | | |
|---|---|---|---|---|---|---|---|---|---|
| 1 | 4 | 10 | 3753.98 | -0.09169 | | | | | |
| | | | | | 1 | 3767.00 | 0.00000 | -0.00832 | -0.01538 | 0.00000 |
| | | | | | 2 | 3762.00 | -0.00423 | -0.00999 | -0.00792 | 0.00000 |
| | | | | | 3 | 3762.00 | -0.00423 | -0.00999 | -0.00792 | 0.00000 |
| | | | | | 4 | 3767.00 | 0.00000 | -0.00832 | -0.01538 | 0.00000 |
| 1 | 4 | 11 | 3750.40 | -0.09980 | | | | | |
| | | | | | 1 | 3763.00 | 0.00000 | -0.00932 | -0.01730 | 0.00000 |
| | | | | | 2 | 3758.00 | -0.00458 | -0.01049 | -0.00821 | 0.00000 |
| | | | | | 3 | 3758.00 | -0.00458 | -0.01049 | -0.00821 | 0.00000 |
| | | | | | 4 | 3763.00 | 0.00000 | -0.00932 | -0.01730 | 0.00000 |
| 1 | 3 | 11 | 3749.79 | -0.10722 | | | | | |
| | | | | | 1 | 3762.00 | 0.00000 | -0.01024 | -0.01906 | 0.00000 |
| | | | | | 2 | 3757.00 | -0.00490 | -0.01094 | -0.00848 | 0.00000 |
| | | | | | 3 | 3757.00 | -0.00490 | -0.01094 | -0.00848 | 0.00000 |
| | | | | | 4 | 3762.00 | 0.00000 | -0.01024 | -0.01906 | 0.00000 |
| 1 | 3 | 12 | 3746.81 | -0.10768 | | | | | |
| | | | | | 1 | 3759.00 | 0.00000 | -0.01029 | -0.01917 | 0.00000 |
| | | | | | 2 | 3754.00 | -0.00492 | -0.01096 | -0.00849 | 0.00000 |
| | | | | | 3 | 3754.00 | -0.00492 | -0.01096 | -0.00849 | 0.00000 |
| | | | | | 4 | 3759.00 | 0.00000 | -0.01029 | -0.01917 | 0.00000 |
| 1 | 3 | 13 | 3744.07 | -0.11264 | | | | | |
| | | | | | 1 | 3756.00 | 0.00000 | -0.01091 | -0.02034 | 0.00000 |
| | | | | | 2 | 3751.00 | -0.00513 | -0.01126 | -0.00867 | 0.00000 |
| | | | | | 3 | 3751.00 | -0.00513 | -0.01126 | -0.00867 | 0.00000 |
| | | | | | 4 | 3756.00 | 0.00000 | -0.01091 | -0.02034 | 0.00000 |
| 1 | 3 | 14 | 3741.31 | -0.13673 | | | | | |
| | | | | | 1 | 3752.00 | 0.00000 | -0.01386 | -0.02599 | 0.00000 |
| | | | | | 2 | 3747.00 | -0.00717 | -0.01210 | -0.00924 | 0.00000 |
| | | | | | 3 | 3747.00 | -0.00717 | -0.01210 | -0.00924 | 0.00000 |
| | | | | | 4 | 3752.00 | 0.00000 | -0.01386 | -0.02599 | 0.00000 |
| 1 | 3 | 15 | 3738.73 | -0.15909 | | | | | |
| | | | | | 1 | 3748.00 | 0.00000 | -0.01722 | -0.03162 | 0.00000 |
| | | | | | 2 | 3743.00 | -0.00957 | -0.01198 | -0.00915 | 0.00000 |
| | | | | | 3 | 3743.00 | -0.00957 | -0.01198 | -0.00915 | 0.00000 |
| | | | | | 4 | 3748.00 | 0.00000 | -0.01722 | -0.03162 | 0.00000 |
| 1 | 3 | 16 | 3736.72 | -0.14462 | | | | | |
| | | | | | 1 | 3747.00 | 0.00000 | -0.01482 | -0.02783 | 0.00000 |
| | | | | | 2 | 3742.00 | -0.00798 | -0.01229 | -0.00939 | 0.00000 |
| | | | | | 3 | 3742.00 | -0.00798 | -0.01229 | -0.00939 | 0.00000 |
| | | | | | 4 | 3747.00 | 0.00000 | -0.01482 | -0.02783 | 0.00000 |
| 1 | 4 | 16 | 3735.61 | -0.14258 | | | | | |
| | | | | | 1 | 3746.00 | 0.00000 | -0.01457 | -0.02736 | 0.00000 |
| | | | | | 2 | 3741.00 | -0.00777 | -0.01224 | -0.00935 | 0.00000 |
| | | | | | 3 | 3741.00 | -0.00777 | -0.01224 | -0.00935 | 0.00000 |
| | | | | | 4 | 3746.00 | 0.00000 | -0.01457 | -0.02736 | 0.00000 |
| 1 | 5 | 16 | 3734.28 | -0.13621 | | | | | |
| | | | | | 1 | 3745.00 | 0.00000 | -0.01379 | -0.02587 | 0.00000 |
| | | | | | 2 | 3740.00 | -0.00712 | -0.01209 | -0.00923 | 0.00000 |
| | | | | | 3 | 3740.00 | -0.00712 | -0.01209 | -0.00923 | 0.00000 |
| | | | | | 4 | 3745.00 | 0.00000 | -0.01379 | -0.02587 | 0.00000 |
| 1 | 6 | 16 | 3733.04 | -0.13144 | | | | | |
| | | | | | 1 | 3744.00 | 0.00000 | -0.01321 | -0.02476 | 0.00000 |
| | | | | | 2 | 3739.00 | -0.00663 | -0.01198 | -0.00914 | 0.00000 |
| | | | | | 3 | 3739.00 | -0.00663 | -0.01198 | -0.00914 | 0.00000 |
| | | | | | 4 | 3744.00 | 0.00000 | -0.01321 | -0.02476 | 0.00000 |
| 1 | 6 | 17 | 3729.35 | -0.11797 | | | | | |
| | | | | | 1 | 3741.00 | 0.00000 | -0.01157 | -0.02161 | 0.00000 |
| | | | | | 2 | 3736.00 | -0.00536 | -0.01159 | -0.00886 | 0.00000 |
| | | | | | 3 | 3736.00 | -0.00536 | -0.01159 | -0.00886 | 0.00000 |
| | | | | | 4 | 3741.00 | 0.00000 | -0.01157 | -0.02161 | 0.00000 |
| 1 | 6 | 18 | 3725.16 | -0.11432 | | | | | |
| | | | | | 1 | 3737.00 | 0.00000 | -0.01112 | -0.02074 | 0.00000 |
| | | | | | 2 | 3732.00 | -0.00520 | -0.01137 | -0.00873 | 0.00000 |
| | | | | | 3 | 3732.00 | -0.00520 | -0.01137 | -0.00873 | 0.00000 |
| | | | | | 4 | 3737.00 | 0.00000 | -0.01112 | -0.02074 | 0.00000 |
| 1 | 6 | 19 | 3720.38 | -0.08007 | | | | | |
| | | | | | 1 | 3734.00 | 0.00000 | -0.00685 | -0.01265 | 0.00000 |
| | | | | | 2 | 3729.00 | -0.00373 | -0.00930 | -0.00751 | 0.00000 |
| | | | | | 3 | 3729.00 | -0.00373 | -0.00930 | -0.00751 | 0.00000 |
| | | | | | 4 | 3734.00 | 0.00000 | -0.00685 | -0.01265 | 0.00000 |
| 1 | 6 | 20 | 3715.59 | -0.06482 | | | | | |
| | | | | | 1 | 3730.00 | 0.00000 | -0.00491 | -0.00907 | 0.00000 |
| | | | | | 2 | 3725.00 | -0.00308 | -0.00838 | -0.00697 | 0.00000 |
| | | | | | 3 | 3725.00 | -0.00308 | -0.00838 | -0.00697 | 0.00000 |
| | | | | | 4 | 3730.00 | 0.00000 | -0.00491 | -0.00907 | 0.00000 |
| | | Total ET per PFSG | | | | | -0.27667 | -1.11362 | -1.50385 | -0.12587 |

| LAYER | ROW | COLUMN | STREAM NUMBER | REACH NUMBER | FLOW INTO STREAM REACH | FLOW INTO AQUIFER | FLOW OUT OF STREAM REACH | HEAD IN STREAM |
|---|---|---|---|---|---|---|---|---|
| 1 | 2 | 2 | 1 | 1 | 100.0000 | -0.5033950 | 100.6034 | 3786.91 |
| 1 | 2 | 3 | 1 | 2 | 100.6034 | -0.2060276 | 100.7094 | 3783.83 |
| 1 | 2 | 4 | 1 | 3 | 100.7094 | -0.1386451 | 100.8481 | 3780.27 |
| 1 | 3 | 4 | 1 | 4 | 100.8481 | 0.4818515E-01 | 100.7999 | 3779.27 |
| 1 | 4 | 4 | 1 | 5 | 100.7999 | 0.2892914 | 100.5106 | 3778.03 |
| 1 | 4 | 5 | 1 | 6 | 100.5106 | 0.3629678 | 100.1476 | 3776.02 |
| 1 | 4 | 6 | 1 | 7 | 100.1476 | 0.4844007 | 99.66322 | 3773.78 |
| 1 | 4 | 7 | 1 | 8 | 99.66322 | 0.3782174 | 99.28500 | 3768.83 |
| 1 | 4 | 8 | 1 | 9 | 99.28500 | 0.3476505 | 98.93735 | 3764.83 |
| 1 | 4 | 9 | 1 | 10 | 98.93735 | 0.3159557 | 98.62139 | 3760.90 |
| 1 | 4 | 10 | 1 | 11 | 98.62139 | 0.3551174 | 98.26627 | 3757.82 |
| 1 | 4 | 11 | 1 | 12 | 98.26627 | 0.3551851 | 97.91109 | 3754.24 |
| 1 | 3 | 11 | 1 | 13 | 97.91109 | 0.3192267 | 97.59186 | 3753.24 |
| 1 | 3 | 12 | 1 | 14 | 97.59186 | 0.2845992 | 97.30726 | 3749.89 |
| 1 | 3 | 13 | 1 | 15 | 97.30726 | 0.2605969 | 97.04666 | 3746.89 |
| 1 | 3 | 14 | 1 | 16 | 97.04666 | 0.1390286 | 96.90763 | 3742.81 |
| 1 | 3 | 15 | 1 | 17 | 96.90763 | 0.7466895E-02 | 96.90016 | 3738.81 |
| 1 | 3 | 16 | 1 | 18 | 96.90016 | 0.1404272 | 96.75974 | 3738.24 |
| 1 | 4 | 16 | 1 | 19 | 96.75974 | 0.1500372 | 96.60970 | 3737.23 |
| 1 | 5 | 16 | 1 | 20 | 96.60970 | 0.1800853 | 96.42961 | 3736.23 |
| 1 | 6 | 16 | 1 | 21 | 96.42961 | 0.2025536 | 96.22706 | 3735.23 |
| 1 | 6 | 17 | 1 | 22 | 96.22706 | 0.2344289 | 95.99263 | 3731.88 |
| 1 | 6 | 18 | 1 | 23 | 95.99263 | 0.2450991 | 95.74763 | 3727.81 |
| 1 | 6 | 19 | 1 | 24 | 95.74763 | 0.4159805 | 95.33155 | 3724.88 |
| 1 | 6 | 20 | 1 | 25 | 95.33155 | 0.4816937 | 94.84985 | 3720.81 |

1

HEAD IN LAYER   1 AT END OF TIME STEP   1 IN STRESS PERIOD   1

----------------------------------------------------------------------

|   | 1 | 2 | 3 | 4 | 5 | 6 | 7 | 8 | 9 | 10 | 11 | 12 | 13 | 14 | 15 | 16 | 17 | 18 | 19 | 20 |
|---|---|---|---|---|---|---|---|---|---|---|---|---|---|---|---|---|---|---|---|---|
| 1 | 3800. | 3793. | 3788. | 3784. | 3782. | -999. | -999. | -999. | -999. | -999. | 3746. | 3745. | 3743. | 3742. | 3740. | 3738. | 3737. | 3736. | -999. | -999. |
| 2 | 3800. | 3792. | 3786. | 3782. | 3780. | -999. | -999. | -999. | -999. | -999. | 3747. | 3746. | 3744. | 3741. | 3739. | 3738. | 3736. | 3735. | -999. | -999. |
| 3 | 3800. | 3792. | 3784. | 3779. | 3775. | 3770. | 3765. | 3761. | 3758. | 3754. | 3750. | 3747. | 3744. | 3741. | 3739. | 3737. | 3735. | 3734. | -999. | -999. |
| 4 | -999. | -999. | 3777. | 3775. | 3772. | 3769. | 3765. | 3761. | 3757. | 3754. | 3750. | 3747. | 3744. | 3741. | 3738. | 3736. | 3733. | 3731. | -999. | -999. |
| 5 | -999. | -999. | 3772. | 3771. | 3769. | 3766. | 3763. | 3760. | 3756. | 3753. | 3750. | 3747. | 3743. | 3740. | 3737. | 3734. | 3731. | 3727. | 3721. | 3716. |
| 6 | -999. | -999. | 3769. | 3768. | 3766. | 3764. | 3761. | 3758. | 3755. | 3752. | 3749. | 3746. | 3743. | 3740. | 3736. | 3733. | 3729. | 3725. | 3720. | 3716. |
| 7 | -999. | -999. | 3767. | 3766. | 3764. | 3762. | 3760. | 3757. | 3754. | 3752. | 3749. | 3746. | 3742. | 3739. | 3736. | 3732. | 3728. | 3723. | 3719. | 3714. |
| 8 | -999. | -999. | 3765. | 3764. | 3762. | 3760. | 3758. | 3756. | 3753. | 3751. | 3748. | 3745. | 3742. | 3739. | 3735. | 3731. | 3727. | 3722. | 3717. | 3712. |
| 9 | -999. | -999. | 3764. | 3763. | 3761. | 3759. | 3757. | 3755. | 3753. | 3750. | 3748. | 3745. | 3742. | 3739. | 3735. | 3730. | 3726. | 3721. | 3716. | 3711. |
| 10 | -999. | -999. | -999. | -999. | -999. | 3756. | 3754. | 3752. | 3750. | 3748. | 3745. | 3743. | 3740. | 3735. | 3730. | 3726. | 3721. | 3716. | 3711. |
| 11 | -999. | -999. | -999. | -999. | -999. | 3755. | 3754. | 3753. | 3751. | 3750. | 3748. | 3746. | 3744. | 3742. | -999. | -999. | -999. | -999. | -999. | -999. |
| 12 | -999. | -999. | -999. | -999. | -999. | 3754. | 3754. | 3753. | 3751. | 3749. | 3748. | 3746. | 3744. | 3743. | -999. | -999. | -999. | -999. | -999. | -999. |

HEAD WILL BE SAVED ON UNIT   30 AT END OF TIME STEP   1, STRESS PERIOD   1

1

```
    VOLUMETRIC BUDGET FOR ENTIRE MODEL AT END OF TIME STEP   1 IN STRESS PERIOD    1
    ----------------------------------------------------------------------------------

        CUMULATIVE VOLUMES     L**3        RATES FOR THIS TIME STEP    L**3/T
        ------------------                 ------------------------

            IN:                                IN:
            ---                                ---
                STORAGE =    3504340.0000          STORAGE =        0.2221
          CONSTANT HEAD =  108875872.0000     CONSTANT HEAD =        6.9001
                  WELLS =          0.0000             WELLS =        0.0000
            RIPARIAN ET =          0.0000       RIPARIAN ET =        0.0000
         STREAM LEAKAGE =   94644312.0000    STREAM LEAKAGE =        5.9982

               TOTAL IN =  207024512.0000          TOTAL IN =       13.1204

           OUT:                               OUT:
           ----                               ----
                STORAGE =   51324560.0000          STORAGE =        3.2528
          CONSTANT HEAD =          0.0000     CONSTANT HEAD =        0.0000
                  WELLS =   94672800.0000             WELLS =        6.0000
            RIPARIAN ET =   47652164.0000       RIPARIAN ET =        3.0200
         STREAM LEAKAGE =   13381492.0000    STREAM LEAKAGE =        0.8481

              TOTAL OUT =  207031008.0000         TOTAL OUT =       13.1208

               IN - OUT =      -6496.0000          IN - OUT =   -4.1199E-04

    PERCENT DISCREPANCY =          0.00   PERCENT DISCREPANCY =        0.00
```

```
       TIME SUMMARY AT END OF TIME STEP   1 IN STRESS PERIOD    1
                   SECONDS      MINUTES      HOURS        DAYS        YEARS
                  ----------------------------------------------------------------
    TIME STEP LENGTH 1.57788E+07 2.62980E+05   4383.0      182.62      0.50000
  STRESS PERIOD TIME 1.57788E+07 2.62980E+05   4383.0      182.62      0.50000
         TOTAL TIME 1.57788E+07 2.62980E+05   4383.0      182.62      0.50000
```
1
1

```
                  STRESS PERIOD NO.    2, LENGTH =  0.1577880E+08
                  ------------------------------------------------

                      NUMBER OF TIME STEPS =     1

                      MULTIPLIER FOR DELT =    1.000

                  INITIAL TIME STEP SIZE =  0.1577880E+08
```
REUSING NON-PARAMETER WELLS FROM LAST STRESS PERIOD

     6 WELLS

25 RIPARIAN CELLS

RIPARIAN CELL INFORMATION

| LAYER | ROW | COLUMN | POLYGON NUMBER | SURFACE ELEVATION | FRACTIONAL COVERAGE OF PLANT FUNCTIONAL SUBGROUPS | | | |
|---|---|---|---|---|---|---|---|---|
| | | | | | D.R. Riparian Small | D.R. Riparian Medium | D.R. Riparian Large | Evaporation |
| 1 | 2 | 2 | 1 | 3796.00 | 0.00000 | 0.00000 | 0.00000 | 0.03788 |
| | | | 2 | 3791.00 | 0.00000 | 0.00000 | 0.00000 | 0.01894 |
| | | | 3 | 3791.00 | 0.00000 | 0.00000 | 0.00000 | 0.01894 |
| | | | 4 | 3796.00 | 0.00000 | 0.00000 | 0.00000 | 0.03788 |
| 1 | 2 | 3 | 1 | 3793.00 | 0.00000 | 0.00000 | 0.00000 | 0.03788 |
| | | | 2 | 3788.00 | 0.00000 | 0.00000 | 0.00000 | 0.01894 |
| | | | 3 | 3788.00 | 0.00000 | 0.00000 | 0.00000 | 0.01894 |
| | | | 4 | 3793.00 | 0.00000 | 0.00000 | 0.00000 | 0.03788 |
| 1 | 2 | 4 | 1 | 3789.00 | 0.00000 | 0.00000 | 0.00000 | 0.03788 |
| | | | 2 | 3784.00 | 0.00000 | 0.00000 | 0.00000 | 0.01894 |
| | | | 3 | 3784.00 | 0.00000 | 0.00000 | 0.00000 | 0.01894 |
| | | | 4 | 3789.00 | 0.00000 | 0.00000 | 0.00000 | 0.03788 |
| 1 | 3 | 4 | 1 | 3788.00 | 0.00000 | 0.00000 | 0.00000 | 0.03788 |
| | | | 2 | 3783.00 | 0.00000 | 0.00000 | 0.00000 | 0.01894 |
| | | | 3 | 3783.00 | 0.00000 | 0.00000 | 0.00000 | 0.01894 |
| | | | 4 | 3788.00 | 0.00000 | 0.00000 | 0.00000 | 0.03788 |
| 1 | 4 | 4 | 1 | 3787.00 | 0.00000 | 0.00000 | 0.00000 | 0.03788 |
| | | | 2 | 3782.00 | 0.00000 | 0.00000 | 0.00000 | 0.01894 |
| | | | 3 | 3782.00 | 0.00000 | 0.00000 | 0.00000 | 0.01894 |
| | | | 4 | 3787.00 | 0.00000 | 0.00000 | 0.00000 | 0.03788 |
| 1 | 4 | 5 | 1 | 3785.00 | 0.00000 | 0.00000 | 0.00000 | 0.03788 |
| | | | 2 | 3780.00 | 0.00000 | 0.00000 | 0.00000 | 0.01894 |
| | | | 3 | 3780.00 | 0.00000 | 0.00000 | 0.00000 | 0.01894 |
| | | | 4 | 3785.00 | 0.00000 | 0.00000 | 0.00000 | 0.03788 |
| 1 | 4 | 6 | 1 | 3783.00 | 0.00000 | 0.00000 | 0.00000 | 0.03788 |
| | | | 2 | 3778.00 | 0.00000 | 0.00000 | 0.00000 | 0.01894 |
| | | | 3 | 3778.00 | 0.00000 | 0.00000 | 0.00000 | 0.01894 |
| | | | 4 | 3783.00 | 0.00000 | 0.00000 | 0.00000 | 0.03788 |
| 1 | 4 | 7 | 1 | 3778.00 | 0.00000 | 0.00000 | 0.00000 | 0.03788 |
| | | | 2 | 3773.00 | 0.00000 | 0.00000 | 0.00000 | 0.01894 |
| | | | 3 | 3773.00 | 0.00000 | 0.00000 | 0.00000 | 0.01894 |
| | | | 4 | 3778.00 | 0.00000 | 0.00000 | 0.00000 | 0.03788 |
| 1 | 4 | 8 | 1 | 3774.00 | 0.00000 | 0.00000 | 0.00000 | 0.03788 |
| | | | 2 | 3769.00 | 0.00000 | 0.00000 | 0.00000 | 0.01894 |
| | | | 3 | 3769.00 | 0.00000 | 0.00000 | 0.00000 | 0.01894 |
| | | | 4 | 3774.00 | 0.00000 | 0.00000 | 0.00000 | 0.03788 |
| 1 | 4 | 9 | 1 | 3770.00 | 0.00000 | 0.00000 | 0.00000 | 0.03788 |
| | | | 2 | 3765.00 | 0.00000 | 0.00000 | 0.00000 | 0.01894 |
| | | | 3 | 3765.00 | 0.00000 | 0.00000 | 0.00000 | 0.01894 |
| | | | 4 | 3770.00 | 0.00000 | 0.00000 | 0.00000 | 0.03788 |
| 1 | 4 | 10 | 1 | 3767.00 | 0.00000 | 0.00000 | 0.00000 | 0.03788 |
| | | | 2 | 3762.00 | 0.00000 | 0.00000 | 0.00000 | 0.01894 |
| | | | 3 | 3762.00 | 0.00000 | 0.00000 | 0.00000 | 0.01894 |
| | | | 4 | 3767.00 | 0.00000 | 0.00000 | 0.00000 | 0.03788 |
| 1 | 4 | 11 | 1 | 3763.00 | 0.00000 | 0.00000 | 0.00000 | 0.03788 |
| | | | 2 | 3758.00 | 0.00000 | 0.00000 | 0.00000 | 0.01894 |
| | | | 3 | 3758.00 | 0.00000 | 0.00000 | 0.00000 | 0.01894 |
| | | | 4 | 3763.00 | 0.00000 | 0.00000 | 0.00000 | 0.03788 |
| 1 | 3 | 11 | 1 | 3762.00 | 0.00000 | 0.00000 | 0.00000 | 0.03788 |
| | | | 2 | 3757.00 | 0.00000 | 0.00000 | 0.00000 | 0.01894 |
| | | | 3 | 3757.00 | 0.00000 | 0.00000 | 0.00000 | 0.01894 |
| | | | 4 | 3762.00 | 0.00000 | 0.00000 | 0.00000 | 0.03788 |
| 1 | 3 | 12 | 1 | 3759.00 | 0.00000 | 0.00000 | 0.00000 | 0.03788 |
| | | | 2 | 3754.00 | 0.00000 | 0.00000 | 0.00000 | 0.01894 |
| | | | 3 | 3754.00 | 0.00000 | 0.00000 | 0.00000 | 0.01894 |
| | | | 4 | 3759.00 | 0.00000 | 0.00000 | 0.00000 | 0.03788 |
| 1 | 3 | 13 | 1 | 3756.00 | 0.00000 | 0.00000 | 0.00000 | 0.03788 |
| | | | 2 | 3751.00 | 0.00000 | 0.00000 | 0.00000 | 0.01894 |
| | | | 3 | 3751.00 | 0.00000 | 0.00000 | 0.00000 | 0.01894 |
| | | | 4 | 3756.00 | 0.00000 | 0.00000 | 0.00000 | 0.03788 |
| 1 | 3 | 14 | 1 | 3752.00 | 0.00000 | 0.00000 | 0.00000 | 0.03788 |
| | | | 2 | 3747.00 | 0.00000 | 0.00000 | 0.00000 | 0.01894 |
| | | | 3 | 3747.00 | 0.00000 | 0.00000 | 0.00000 | 0.01894 |
| | | | 4 | 3752.00 | 0.00000 | 0.00000 | 0.00000 | 0.03788 |
| 1 | 3 | 15 | 1 | 3748.00 | 0.00000 | 0.00000 | 0.00000 | 0.03788 |
| | | | 2 | 3743.00 | 0.00000 | 0.00000 | 0.00000 | 0.01894 |
| | | | 3 | 3743.00 | 0.00000 | 0.00000 | 0.00000 | 0.01894 |
| | | | 4 | 3748.00 | 0.00000 | 0.00000 | 0.00000 | 0.03788 |

| | | | | | | | | | |
|---|---|---|---|---|---|---|---|---|---|
| 1 | 3 | 16 | 1 | 3747.00 | 0.00000 | 0.00000 | 0.00000 | 0.03788 |
| | | | 2 | 3742.00 | 0.00000 | 0.00000 | 0.00000 | 0.01894 |
| | | | 3 | 3742.00 | 0.00000 | 0.00000 | 0.00000 | 0.01894 |
| | | | 4 | 3747.00 | 0.00000 | 0.00000 | 0.00000 | 0.03788 |
| 1 | 4 | 16 | 1 | 3746.00 | 0.00000 | 0.00000 | 0.00000 | 0.03788 |
| | | | 2 | 3741.00 | 0.00000 | 0.00000 | 0.00000 | 0.01894 |
| | | | 3 | 3741.00 | 0.00000 | 0.00000 | 0.00000 | 0.01894 |
| | | | 4 | 3746.00 | 0.00000 | 0.00000 | 0.00000 | 0.03788 |
| 1 | 5 | 16 | 1 | 3745.00 | 0.00000 | 0.00000 | 0.00000 | 0.03788 |
| | | | 2 | 3740.00 | 0.00000 | 0.00000 | 0.00000 | 0.01894 |
| | | | 3 | 3740.00 | 0.00000 | 0.00000 | 0.00000 | 0.01894 |
| | | | 4 | 3745.00 | 0.00000 | 0.00000 | 0.00000 | 0.03788 |
| 1 | 6 | 16 | 1 | 3744.00 | 0.00000 | 0.00000 | 0.00000 | 0.03788 |
| | | | 2 | 3739.00 | 0.00000 | 0.00000 | 0.00000 | 0.01894 |
| | | | 3 | 3739.00 | 0.00000 | 0.00000 | 0.00000 | 0.01894 |
| | | | 4 | 3744.00 | 0.00000 | 0.00000 | 0.00000 | 0.03788 |
| 1 | 6 | 17 | 1 | 3741.00 | 0.00000 | 0.00000 | 0.00000 | 0.03788 |
| | | | 2 | 3736.00 | 0.00000 | 0.00000 | 0.00000 | 0.01894 |
| | | | 3 | 3736.00 | 0.00000 | 0.00000 | 0.00000 | 0.01894 |
| | | | 4 | 3741.00 | 0.00000 | 0.00000 | 0.00000 | 0.03788 |
| 1 | 6 | 18 | 1 | 3737.00 | 0.00000 | 0.00000 | 0.00000 | 0.03788 |
| | | | 2 | 3732.00 | 0.00000 | 0.00000 | 0.00000 | 0.01894 |
| | | | 3 | 3732.00 | 0.00000 | 0.00000 | 0.00000 | 0.01894 |
| | | | 4 | 3737.00 | 0.00000 | 0.00000 | 0.00000 | 0.03788 |
| 1 | 6 | 19 | 1 | 3734.00 | 0.00000 | 0.00000 | 0.00000 | 0.03788 |
| | | | 2 | 3729.00 | 0.00000 | 0.00000 | 0.00000 | 0.01894 |
| | | | 3 | 3729.00 | 0.00000 | 0.00000 | 0.00000 | 0.01894 |
| | | | 4 | 3734.00 | 0.00000 | 0.00000 | 0.00000 | 0.03788 |
| 1 | 6 | 20 | 1 | 3730.00 | 0.00000 | 0.00000 | 0.00000 | 0.03788 |
| | | | 2 | 3725.00 | 0.00000 | 0.00000 | 0.00000 | 0.01894 |
| | | | 3 | 3725.00 | 0.00000 | 0.00000 | 0.00000 | 0.01894 |
| | | | 4 | 3730.00 | 0.00000 | 0.00000 | 0.00000 | 0.03788 |

REUSING STREAM NODES FROM LAST STRESS PERIOD

SOLVING FOR HEAD

   14 ITERATIONS FOR TIME STEP   1 IN STRESS PERIOD    2

MAXIMUM HEAD CHANGE FOR EACH ITERATION:

```
   HEAD CHANGE    HEAD CHANGE    HEAD CHANGE    HEAD CHANGE    HEAD CHANGE
 LAYER,ROW,COL  LAYER,ROW,COL  LAYER,ROW,COL  LAYER,ROW,COL  LAYER,ROW,COL
-----------------------------------------------------------------------------
     1.892          1.250          1.216          0.4078        0.9720E-01
 (  1, 10, 20)  (  1,  9, 19)  (  1, 10, 20)  (  1, 10, 15)  (  1,  5, 19)
   0.2203E-01     0.1024E-01     0.1174E-01     0.4580E-02     0.2517E-02
 (  1, 10, 15)  (  1, 10, 16)  (  1, 11, 14)  (  1, 10, 20)  (  1, 11, 14)
  -0.3013E-03     0.1384E-03     0.1263E-03     0.6380E-04
 (  1, 10, 17)  (  1, 12, 13)  (  1, 12, 13)  (  1, 12,  6)
```

OUTPUT CONTROL FOR STRESS PERIOD    2  TIME STEP   1
     PRINT HEAD FOR ALL LAYERS
     PRINT BUDGET
     SAVE BUDGET
     SAVE HEAD FOR ALL LAYERS

     CONSTANT HEAD   PERIOD  2   STEP  1
 LAYER  1   ROW   1   COL   1   RATE   2.02026
 LAYER  1   ROW   2   COL   1   RATE   2.34970
 LAYER  1   ROW   3   COL   1   RATE   2.43825

          WELLS   PERIOD   2   STEP   1
 WELL    1  LAYER  1  ROW    5  COL  20  RATE  -1.00000
 WELL    2  LAYER  1  ROW    6  COL  20  RATE  -1.00000
 WELL    3  LAYER  1  ROW    7  COL  20  RATE  -1.00000
 WELL    4  LAYER  1  ROW    8  COL  20  RATE  -1.00000
 WELL    5  LAYER  1  ROW    9  COL  20  RATE  -1.00000
 WELL    6  LAYER  1  ROW   10  COL  20  RATE  -1.00000

RIPARIAN ET    PERIOD  2    STEP  1

| RIPARIAN ET LAYER | ROW | COLUMN | HEAD | CELL ET RATE | POLYGON NUMBER | SURFACE ELEVATION | CELL ET RATE BY PLANT FUNCTIONAL GROUP D.R. Riparian Small | D.R. Riparian Medium | D.R. Riparian Large | Evaporation |
|---|---|---|---|---|---|---|---|---|---|---|
| 1 | 2 | 2 | 3792.44 | -0.22071 | | | | | | |
| | | | | | 1 | 3796.00 | 0.00000 | 0.00000 | 0.00000 | 0.00000 |
| | | | | | 2 | 3791.00 | 0.00000 | 0.00000 | 0.00000 | -0.11036 |
| | | | | | 3 | 3791.00 | 0.00000 | 0.00000 | 0.00000 | -0.11036 |
| | | | | | 4 | 3796.00 | 0.00000 | 0.00000 | 0.00000 | 0.00000 |
| 1 | 2 | 3 | 3786.33 | -0.10855 | | | | | | |
| | | | | | 1 | 3793.00 | 0.00000 | 0.00000 | 0.00000 | 0.00000 |
| | | | | | 2 | 3788.00 | 0.00000 | 0.00000 | 0.00000 | -0.05428 |
| | | | | | 3 | 3788.00 | 0.00000 | 0.00000 | 0.00000 | -0.05428 |
| | | | | | 4 | 3793.00 | 0.00000 | 0.00000 | 0.00000 | 0.00000 |
| 1 | 2 | 4 | 3782.14 | -0.09525 | | | | | | |
| | | | | | 1 | 3789.00 | 0.00000 | 0.00000 | 0.00000 | 0.00000 |
| | | | | | 2 | 3784.00 | 0.00000 | 0.00000 | 0.00000 | -0.04763 |
| | | | | | 3 | 3784.00 | 0.00000 | 0.00000 | 0.00000 | -0.04763 |
| | | | | | 4 | 3789.00 | 0.00000 | 0.00000 | 0.00000 | 0.00000 |
| 1 | 3 | 4 | 3779.20 | 0.00000 | | | | | | |
| | | | | | 1 | 3788.00 | 0.00000 | 0.00000 | 0.00000 | 0.00000 |
| | | | | | 2 | 3783.00 | 0.00000 | 0.00000 | 0.00000 | 0.00000 |
| | | | | | 3 | 3783.00 | 0.00000 | 0.00000 | 0.00000 | 0.00000 |
| | | | | | 4 | 3788.00 | 0.00000 | 0.00000 | 0.00000 | 0.00000 |
| 1 | 4 | 4 | 3775.33 | 0.00000 | | | | | | |
| | | | | | 1 | 3787.00 | 0.00000 | 0.00000 | 0.00000 | 0.00000 |
| | | | | | 2 | 3782.00 | 0.00000 | 0.00000 | 0.00000 | 0.00000 |
| | | | | | 3 | 3782.00 | 0.00000 | 0.00000 | 0.00000 | 0.00000 |
| | | | | | 4 | 3787.00 | 0.00000 | 0.00000 | 0.00000 | 0.00000 |
| 1 | 4 | 5 | 3772.54 | 0.00000 | | | | | | |
| | | | | | 1 | 3785.00 | 0.00000 | 0.00000 | 0.00000 | 0.00000 |
| | | | | | 2 | 3780.00 | 0.00000 | 0.00000 | 0.00000 | 0.00000 |
| | | | | | 3 | 3780.00 | 0.00000 | 0.00000 | 0.00000 | 0.00000 |
| | | | | | 4 | 3785.00 | 0.00000 | 0.00000 | 0.00000 | 0.00000 |
| 1 | 4 | 6 | 3768.97 | 0.00000 | | | | | | |
| | | | | | 1 | 3783.00 | 0.00000 | 0.00000 | 0.00000 | 0.00000 |
| | | | | | 2 | 3778.00 | 0.00000 | 0.00000 | 0.00000 | 0.00000 |
| | | | | | 3 | 3778.00 | 0.00000 | 0.00000 | 0.00000 | 0.00000 |
| | | | | | 4 | 3783.00 | 0.00000 | 0.00000 | 0.00000 | 0.00000 |
| 1 | 4 | 7 | 3765.21 | 0.00000 | | | | | | |
| | | | | | 1 | 3778.00 | 0.00000 | 0.00000 | 0.00000 | 0.00000 |
| | | | | | 2 | 3773.00 | 0.00000 | 0.00000 | 0.00000 | 0.00000 |
| | | | | | 3 | 3773.00 | 0.00000 | 0.00000 | 0.00000 | 0.00000 |
| | | | | | 4 | 3778.00 | 0.00000 | 0.00000 | 0.00000 | 0.00000 |
| 1 | 4 | 8 | 3761.59 | 0.00000 | | | | | | |
| | | | | | 1 | 3774.00 | 0.00000 | 0.00000 | 0.00000 | 0.00000 |
| | | | | | 2 | 3769.00 | 0.00000 | 0.00000 | 0.00000 | 0.00000 |
| | | | | | 3 | 3769.00 | 0.00000 | 0.00000 | 0.00000 | 0.00000 |
| | | | | | 4 | 3774.00 | 0.00000 | 0.00000 | 0.00000 | 0.00000 |
| 1 | 4 | 9 | 3758.06 | 0.00000 | | | | | | |
| | | | | | 1 | 3770.00 | 0.00000 | 0.00000 | 0.00000 | 0.00000 |
| | | | | | 2 | 3765.00 | 0.00000 | 0.00000 | 0.00000 | 0.00000 |
| | | | | | 3 | 3765.00 | 0.00000 | 0.00000 | 0.00000 | 0.00000 |
| | | | | | 4 | 3770.00 | 0.00000 | 0.00000 | 0.00000 | 0.00000 |
| 1 | 4 | 10 | 3754.61 | 0.00000 | | | | | | |
| | | | | | 1 | 3767.00 | 0.00000 | 0.00000 | 0.00000 | 0.00000 |
| | | | | | 2 | 3762.00 | 0.00000 | 0.00000 | 0.00000 | 0.00000 |
| | | | | | 3 | 3762.00 | 0.00000 | 0.00000 | 0.00000 | 0.00000 |
| | | | | | 4 | 3767.00 | 0.00000 | 0.00000 | 0.00000 | 0.00000 |
| 1 | 4 | 11 | 3751.10 | 0.00000 | | | | | | |
| | | | | | 1 | 3763.00 | 0.00000 | 0.00000 | 0.00000 | 0.00000 |
| | | | | | 2 | 3758.00 | 0.00000 | 0.00000 | 0.00000 | 0.00000 |
| | | | | | 3 | 3758.00 | 0.00000 | 0.00000 | 0.00000 | 0.00000 |
| | | | | | 4 | 3763.00 | 0.00000 | 0.00000 | 0.00000 | 0.00000 |
| 1 | 3 | 11 | 3750.50 | 0.00000 | | | | | | |
| | | | | | 1 | 3762.00 | 0.00000 | 0.00000 | 0.00000 | 0.00000 |
| | | | | | 2 | 3757.00 | 0.00000 | 0.00000 | 0.00000 | 0.00000 |
| | | | | | 3 | 3757.00 | 0.00000 | 0.00000 | 0.00000 | 0.00000 |
| | | | | | 4 | 3762.00 | 0.00000 | 0.00000 | 0.00000 | 0.00000 |
| 1 | 3 | 12 | 3747.58 | 0.00000 | | | | | | |
| | | | | | 1 | 3759.00 | 0.00000 | 0.00000 | 0.00000 | 0.00000 |
| | | | | | 2 | 3754.00 | 0.00000 | 0.00000 | 0.00000 | 0.00000 |
| | | | | | 3 | 3754.00 | 0.00000 | 0.00000 | 0.00000 | 0.00000 |
| | | | | | 4 | 3759.00 | 0.00000 | 0.00000 | 0.00000 | 0.00000 |
| 1 | 3 | 13 | 3744.92 | 0.00000 | | | | | | |
| | | | | | 1 | 3756.00 | 0.00000 | 0.00000 | 0.00000 | 0.00000 |
| | | | | | 2 | 3751.00 | 0.00000 | 0.00000 | 0.00000 | 0.00000 |
| | | | | | 3 | 3751.00 | 0.00000 | 0.00000 | 0.00000 | 0.00000 |
| | | | | | 4 | 3756.00 | 0.00000 | 0.00000 | 0.00000 | 0.00000 |
| 1 | 3 | 14 | 3742.28 | 0.00000 | | | | | | |
| | | | | | 1 | 3752.00 | 0.00000 | 0.00000 | 0.00000 | 0.00000 |
| | | | | | 2 | 3747.00 | 0.00000 | 0.00000 | 0.00000 | 0.00000 |
| | | | | | 3 | 3747.00 | 0.00000 | 0.00000 | 0.00000 | 0.00000 |
| | | | | | 4 | 3752.00 | 0.00000 | 0.00000 | 0.00000 | 0.00000 |
| 1 | 3 | 15 | 3739.82 | -0.00669 | | | | | | |
| | | | | | 1 | 3748.00 | 0.00000 | 0.00000 | 0.00000 | 0.00000 |
| | | | | | 2 | 3743.00 | 0.00000 | 0.00000 | 0.00000 | -0.00334 |
| | | | | | 3 | 3743.00 | 0.00000 | 0.00000 | 0.00000 | -0.00334 |
| | | | | | 4 | 3748.00 | 0.00000 | 0.00000 | 0.00000 | 0.00000 |
| 1 | 3 | 16 | 3737.89 | 0.00000 | | | | | | |
| | | | | | 1 | 3747.00 | 0.00000 | 0.00000 | 0.00000 | 0.00000 |
| | | | | | 2 | 3742.00 | 0.00000 | 0.00000 | 0.00000 | 0.00000 |
| | | | | | 3 | 3742.00 | 0.00000 | 0.00000 | 0.00000 | 0.00000 |
| | | | | | 4 | 3747.00 | 0.00000 | 0.00000 | 0.00000 | 0.00000 |

| | | | | | | | | | |
|---|---|---|---|---|---|---|---|---|---|
| 1 | 4 | 16 | 3736.91 | 0.00000 | | | | | |
| | | | | | 1 | 3746.00 | 0.00000 | 0.00000 | 0.00000 | 0.00000 |
| | | | | | 2 | 3741.00 | 0.00000 | 0.00000 | 0.00000 | 0.00000 |
| | | | | | 3 | 3741.00 | 0.00000 | 0.00000 | 0.00000 | 0.00000 |
| | | | | | 4 | 3746.00 | 0.00000 | 0.00000 | 0.00000 | 0.00000 |
| 1 | 5 | 16 | 3735.72 | 0.00000 | | | | | |
| | | | | | 1 | 3745.00 | 0.00000 | 0.00000 | 0.00000 | 0.00000 |
| | | | | | 2 | 3740.00 | 0.00000 | 0.00000 | 0.00000 | 0.00000 |
| | | | | | 3 | 3740.00 | 0.00000 | 0.00000 | 0.00000 | 0.00000 |
| | | | | | 4 | 3745.00 | 0.00000 | 0.00000 | 0.00000 | 0.00000 |
| 1 | 6 | 16 | 3734.63 | 0.00000 | | | | | |
| | | | | | 1 | 3744.00 | 0.00000 | 0.00000 | 0.00000 | 0.00000 |
| | | | | | 2 | 3739.00 | 0.00000 | 0.00000 | 0.00000 | 0.00000 |
| | | | | | 3 | 3739.00 | 0.00000 | 0.00000 | 0.00000 | 0.00000 |
| | | | | | 4 | 3744.00 | 0.00000 | 0.00000 | 0.00000 | 0.00000 |
| 1 | 6 | 17 | 3731.21 | 0.00000 | | | | | |
| | | | | | 1 | 3741.00 | 0.00000 | 0.00000 | 0.00000 | 0.00000 |
| | | | | | 2 | 3736.00 | 0.00000 | 0.00000 | 0.00000 | 0.00000 |
| | | | | | 3 | 3736.00 | 0.00000 | 0.00000 | 0.00000 | 0.00000 |
| | | | | | 4 | 3741.00 | 0.00000 | 0.00000 | 0.00000 | 0.00000 |
| 1 | 6 | 18 | 3727.36 | 0.00000 | | | | | |
| | | | | | 1 | 3737.00 | 0.00000 | 0.00000 | 0.00000 | 0.00000 |
| | | | | | 2 | 3732.00 | 0.00000 | 0.00000 | 0.00000 | 0.00000 |
| | | | | | 3 | 3732.00 | 0.00000 | 0.00000 | 0.00000 | 0.00000 |
| | | | | | 4 | 3737.00 | 0.00000 | 0.00000 | 0.00000 | 0.00000 |
| 1 | 6 | 19 | 3722.97 | 0.00000 | | | | | |
| | | | | | 1 | 3734.00 | 0.00000 | 0.00000 | 0.00000 | 0.00000 |
| | | | | | 2 | 3729.00 | 0.00000 | 0.00000 | 0.00000 | 0.00000 |
| | | | | | 3 | 3729.00 | 0.00000 | 0.00000 | 0.00000 | 0.00000 |
| | | | | | 4 | 3734.00 | 0.00000 | 0.00000 | 0.00000 | 0.00000 |
| 1 | 6 | 20 | 3718.51 | 0.00000 | | | | | |
| | | | | | 1 | 3730.00 | 0.00000 | 0.00000 | 0.00000 | 0.00000 |
| | | | | | 2 | 3725.00 | 0.00000 | 0.00000 | 0.00000 | 0.00000 |
| | | | | | 3 | 3725.00 | 0.00000 | 0.00000 | 0.00000 | 0.00000 |
| | | | | | 4 | 3730.00 | 0.00000 | 0.00000 | 0.00000 | 0.00000 |
| | | Total ET per PFSG | | | | | 0.00000 | 0.00000 | 0.00000 | -0.43120 |

| LAYER | ROW | COLUMN | STREAM NUMBER | REACH NUMBER | FLOW INTO STREAM REACH | FLOW INTO AQUIFER | FLOW OUT OF STREAM REACH | HEAD IN STREAM |
|---|---|---|---|---|---|---|---|---|
| 1 | 2 | 2 | 1 | 1 | 100.0000 | -0.5116966 | 100.5117 | 3786.91 |
| 1 | 2 | 3 | 1 | 2 | 100.5117 | -0.2310000 | 100.7427 | 3783.83 |
| 1 | 2 | 4 | 1 | 3 | 100.7427 | -0.1727763 | 100.9155 | 3780.27 |
| 1 | 3 | 4 | 1 | 4 | 100.9155 | 0.5955469E-02 | 100.9095 | 3779.27 |
| 1 | 4 | 4 | 1 | 5 | 100.9095 | 0.2489792 | 100.6605 | 3778.03 |
| 1 | 4 | 5 | 1 | 6 | 100.6605 | 0.3224300 | 100.3381 | 3776.03 |
| 1 | 4 | 6 | 1 | 7 | 100.3381 | 0.4445171 | 99.89359 | 3773.78 |
| 1 | 4 | 7 | 1 | 8 | 99.89359 | 0.3342958 | 99.55929 | 3768.83 |
| 1 | 4 | 8 | 1 | 9 | 99.55929 | 0.2992398 | 99.26005 | 3764.83 |
| 1 | 4 | 9 | 1 | 10 | 99.26005 | 0.2624918 | 98.99756 | 3760.90 |
| 1 | 4 | 10 | 1 | 11 | 98.99756 | 0.2970065 | 98.70055 | 3757.82 |
| 1 | 4 | 11 | 1 | 12 | 98.70055 | 0.2904870 | 98.41006 | 3754.25 |
| 1 | 3 | 11 | 1 | 13 | 98.41006 | 0.2540549 | 98.15601 | 3753.25 |
| 1 | 3 | 12 | 1 | 14 | 98.15601 | 0.2135171 | 97.94250 | 3749.89 |
| 1 | 3 | 13 | 1 | 15 | 97.94250 | 0.1820027 | 97.76050 | 3746.89 |
| 1 | 3 | 14 | 1 | 16 | 97.76050 | 0.4983193E-01 | 97.71066 | 3742.82 |
| 1 | 3 | 15 | 1 | 17 | 97.71066 | -0.9244512E-01 | 97.80311 | 3738.82 |
| 1 | 3 | 16 | 1 | 18 | 97.80311 | 0.3225879E-01 | 97.77085 | 3738.24 |
| 1 | 4 | 16 | 1 | 19 | 97.77085 | 0.3101807E-01 | 97.73983 | 3737.24 |
| 1 | 5 | 16 | 1 | 20 | 97.73983 | 0.4863633E-01 | 97.69119 | 3736.24 |
| 1 | 6 | 16 | 1 | 21 | 97.69119 | 0.5693789E-01 | 97.63425 | 3735.24 |
| 1 | 6 | 17 | 1 | 22 | 97.63425 | 0.6309639E-01 | 97.57116 | 3731.89 |
| 1 | 6 | 18 | 1 | 23 | 97.57116 | 0.4198154E-01 | 97.52917 | 3727.82 |
| 1 | 6 | 19 | 1 | 24 | 97.52917 | 0.1771752 | 97.35200 | 3724.89 |
| 1 | 6 | 20 | 1 | 25 | 97.35200 | 0.2132464 | 97.13875 | 3720.82 |

1

HEAD IN LAYER   1 AT END OF TIME STEP   1 IN STRESS PERIOD    2

----------------------------------------------------------------------

| | 1 | 2 | 3 | 4 | 5 | 6 | 7 | 8 | 9 | 10 | 11 | 12 | 13 | 14 | 15 | 16 | 17 | 18 | 19 | 20 |
|---|---|---|---|---|---|---|---|---|---|---|---|---|---|---|---|---|---|---|---|---|
| 1 | 3800. | 3794. | 3788. | 3784. | 3782. | -999. | -999. | -999. | -999. | -999. | 3747. | 3746. | 3744. | 3742. | 3740. | 3739. | 3738. | 3737. | -999. | -999. |
| 2 | 3800. | 3792. | 3786. | 3782. | 3780. | -999. | -999. | -999. | -999. | -999. | 3748. | 3746. | 3744. | 3742. | 3740. | 3739. | 3737. | 3736. | -999. | -999. |
| 3 | 3800. | 3792. | 3784. | 3779. | 3775. | 3770. | 3766. | 3762. | 3758. | 3754. | 3750. | 3748. | 3745. | 3742. | 3740. | 3738. | 3736. | 3735. | -999. | -999. |
| 4 | -999. | -999. | 3777. | 3775. | 3773. | 3769. | 3765. | 3762. | 3758. | 3755. | 3751. | 3748. | 3745. | 3742. | 3739. | 3737. | 3736. | 3733. | -999. | -999. |
| 5 | -999. | -999. | 3773. | 3771. | 3769. | 3766. | 3763. | 3760. | 3757. | 3754. | 3751. | 3747. | 3744. | 3741. | 3739. | 3736. | 3733. | 3729. | 3724. | 3719. |
| 6 | -999. | -999. | 3769. | 3768. | 3767. | 3764. | 3762. | 3759. | 3756. | 3753. | 3750. | 3747. | 3744. | 3741. | 3738. | 3736. | 3731. | 3727. | 3723. | 3719. |
| 7 | -999. | -999. | 3767. | 3766. | 3764. | 3762. | 3760. | 3757. | 3755. | 3752. | 3749. | 3746. | 3743. | 3740. | 3737. | 3734. | 3730. | 3726. | 3722. | 3717. |
| 8 | -999. | -999. | 3765. | 3764. | 3763. | 3761. | 3758. | 3756. | 3754. | 3751. | 3749. | 3746. | 3743. | 3740. | 3737. | 3733. | 3729. | 3725. | 3721. | 3716. |
| 9 | -999. | -999. | 3764. | 3763. | 3762. | 3759. | 3757. | 3755. | 3753. | 3751. | 3748. | 3746. | 3743. | 3740. | 3737. | 3733. | 3729. | 3724. | 3720. | 3716. |
| 10 | -999. | -999. | -999. | -999. | -999. | 3757. | 3756. | 3754. | 3752. | 3750. | 3748. | 3746. | 3744. | 3741. | 3737. | 3733. | 3728. | 3724. | 3720. | 3716. |
| 11 | -999. | -999. | -999. | -999. | -999. | 3755. | 3755. | 3753. | 3752. | 3750. | 3748. | 3746. | 3744. | 3743. | -999. | -999. | -999. | -999. | -999. | -999. |
| 12 | -999. | -999. | -999. | -999. | -999. | 3755. | 3754. | 3753. | 3752. | 3750. | 3748. | 3746. | 3745. | 3744. | -999. | -999. | -999. | -999. | -999. | -999. |

HEAD WILL BE SAVED ON UNIT   30 AT END OF TIME STEP   1, STRESS PERIOD    2

1
```
VOLUMETRIC BUDGET FOR ENTIRE MODEL AT END OF TIME STEP  1 IN STRESS PERIOD   2
------------------------------------------------------------------------------

     CUMULATIVE VOLUMES      L**3        RATES FOR THIS TIME STEP    L**3/T
     ------------------                  ------------------------

          IN:                                 IN:
          ---                                 ---
            STORAGE =     3504340.0000          STORAGE =         0.0000
      CONSTANT HEAD =   216301168.0000    CONSTANT HEAD =         6.8082
              WELLS =           0.0000            WELLS =         0.0000
        RIPARIAN ET =           0.0000      RIPARIAN ET =         0.0000
     STREAM LEAKAGE =   155695008.0000   STREAM LEAKAGE =         3.8692

           TOTAL IN =   375600512.0000         TOTAL IN =        10.6774

          OUT:                                OUT:
          ----                                ----
            STORAGE =   102422432.0000          STORAGE =         3.2384
      CONSTANT HEAD =           0.0000    CONSTANT HEAD =         0.0000
              WELLS =   189345600.0000            WELLS =         6.0000
        RIPARIAN ET =    54456028.0000      RIPARIAN ET =         0.4312
     STREAM LEAKAGE =    29285228.0000   STREAM LEAKAGE =         1.0079

          TOTAL OUT =   375509280.0000        TOTAL OUT =        10.6775

            IN - OUT =       -8768.0000          IN - OUT =    -1.4496E-04

PERCENT DISCREPANCY =           0.00   PERCENT DISCREPANCY =         0.00



     TIME SUMMARY AT END OF TIME STEP  1 IN STRESS PERIOD   2
                      SECONDS     MINUTES      HOURS       DAYS        YEARS
                   ----------------------------------------------------------
 TIME STEP LENGTH 1.57788E+07 2.62980E+05   4383.0     182.62      0.50000
STRESS PERIOD TIME 1.57788E+07 2.62980E+05  4383.0     182.62      0.50000
      TOTAL TIME 3.15576E+07 5.25960E+05    8766.0     365.25      1.0000
1

Run end date and time (yyyy/mm/dd hh:mm:ss): 2011/06/23  7:36:32
Elapsed run time:  0.124 Seconds
```

# 15.0  Code Documentation for the RIP-ET Package

The RIP-ET Package is composed of one module, seven subroutines, and two functions. The seven subroutines are primary subroutines that are part of MODFLOW 2005 (Harbaugh, 2005). The module, subroutines and functions are in source-code file GWF2RIP3.F90. The subroutines are called directly from the main MODFLOW-2005 program unit. The module is called GWFRIPMODULE and creates pointers and a data type for all variables used exclusively by RIP-ET.

The five subroutines are:

| | |
|---|---|
| GWF2RIP3AR | Allocates memory for all data, and reads data for RIP-ET variables that do not vary during a simulation run. |
| GWF2RIP3RP | Reads data for RIP-ET variables that vary over stress periods. |
| GWF2RIP3FM | Formulates terms needed to solve the groundwater flow equation simulating riparian evapotranspiration, and adds them to the head coefficient array (HCOF) and to the right-hand side array (RHS). |
| GWF2RIP3BD | Computes flow rates simulated as evapotranspiration from riparian systems and writes plant functional subgroup evapotranspiration rates or cell-by-cell flow rates if those options are selected. |
| GWF2RIP3DA | Deallocates RIP-ET memory. |
| SGWF2RIP3PNT | Set pointers to RIP-ET data for grid. |
| SGWF2RIP3PSV | Saves pointers to RIP-ET data for grid. |

The two functions are:

| | |
|---|---|
| HCOFtrm | Computes a component of the term added to head coefficient array (HCOF). |
| RHStrm | Computes a component of the term added to the right-hand side array (RHS). |

## 15.1  Changes to Calling Arguments in Main

Statements must be added to mf2005.f, the main program, to call the RIP-ET package. A unique IUNIT number needs to be assigned to the file type RIP. This is done by assigning 'RIP' to an element within the CUNIT data statement of the main program. In MODFLOW 2005, Version 18, IUNIT(6) was available.

## Add 'RIP ' to the 6th position (or whatever IUNIT location is used) in the CUNIT data statement.

```
DATA CUNIT/'BCF6', 'WEL ', 'DRN ', 'RIV ', 'EVT ', 'RIP ', 'GHB ',  ! 7
     &      'RCH ', 'SIP ', 'DE4 ', '    ', 'OC  ', 'PCG ', 'lmg ',  ! 14
     &      'gwt ', 'FHB ', '    ', 'STR ', 'IBS ', 'CHD ', 'HFB6',  ! 21
     &      'lak ', 'LPF ', 'DIS ', '    ', 'PVAL', '    ', '    ',  ! 28
     &      '    ', '    ', 'ZONE', 'MULT', '    ', '    ', '    ',  ! 35
     &      '    ', 'HUF2', '    ', '    ', '    ', '    ', 'GMG ',  ! 42
     &      '    ', 'sfr ', '    ', 'gage', 'LVDA', '    ', 'lmt6',  ! 49
     &      'MNW1', '    ', '    ', 'KDEP', 'sub ', 'uzf ', '    ',  ! 56
     &      44*'    '/
```

## 15.2 Five Calling Statements to RIP-ET from MAIN

The following calling statements are added to MAIN as specified,

1. After the statement,
   ```
   IF(IUNIT(5).GT.0) CALL GWF2EVT7AR(IUNIT(5),IGRID)
   ```
   add the statement,
   ```
   IF(IUNIT(6).GT.0) CALL GWF2RIP3AR(IUNIT(6),IGRID)
   ```

2. After the statement,
   ```
   IF(IUNIT(5).GT.0) CALL GWF2EVT7RP(IUNIT(5),IGRID)
   ```
   add the statement,
   ```
   IF(IUNIT(6).GT.0) CALL GWF2RIP3RP(IUNIT(6),IGRID)
   ```

3. After the statement,
   ```
   IF(IUNIT(5).GT.0) CALL GWF2EVT7FM(IGRID)
   ```
   add the statement,
   ```
   IF(IUNIT(6).GT.0) CALL GWF2RIP3FM(IGRID NDRTNP, NPDRT, IDRT B, IDRTFL, NRFLOW,
   NOPRDT)
   ```

4. After the statement,I
   ```
   F(IUNIT(5).GT.0) CALL GWF2EVT7BD(KKSTP,KKPER,IGRID)
   ```
   add the statement,
   ```
   IF(IUNIT(6).GT.0) CALL GWF2RIP3BD(KKSTP,KKPER,IGRID)
   ```

## 15.3 GWFRIPMODULE

The module GWFRIPMODULE S provides the declaration of allocatable variable for RIP-ET. The listing is as follows.

```
Module GWFRIPMODULE
    Integer, Save, Pointer :: MAXRIP,IRIPCB,IRIPCB1,NRIPCL
    Integer, Save, Pointer :: MAXPOLY,MAXTS,MXSEG
        Integer, Save, Dimension(:),Pointer    :: NuSeg,NPoly
    Integer, Save, Dimension(:,:), Pointer :: CLoc
    Real, Save, Dimension(:), Pointer      :: Sxd,Ard,Rmax,Rsxd,RIPET
    Real, Save, Dimension(:,:), Pointer    :: fdh,fdr,HSurf
    Real, Save, Dimension(:,:,:), Pointer  :: CovPFSG
        Double Precision, Dimension(:,:,:), Pointer    :: C1,C2
    Character(LEN=24), Save, Dimension(:), Pointer :: RIPNM
    TYPE GWFRIPTYPE
      Integer, Pointer        :: MAXRIP,IRIPCB,IRIPCB1,NRIPCL
      Integer, Pointer        :: MAXPOLY,MAXTS,MXSEG
      Integer, Dimension(:), Pointer       :: NuSeg,NPoly
          Integer, Dimension(:,:), Pointer   :: CLoc
      Real, Dimension(:), Pointer        :: Sxd,Ard,Rmax,Rsxd,RIPET
      Real, Dimension(:,:), Pointer      :: fdh,fdr,HSurf
      Real, Dimension(:,:,:), Pointer    :: CovPFSG
          Double Precision, Dimension(:,:,:), Pointer    :: C1,C2
      Character(LEN=24), Dimension(:), Pointer       :: RIPNM
    END TYPE
    TYPE(GWFRIPTYPE), SAVE :: GWFRIPDAT(10)
End Module GWFRIPMODULE
```

## 15.4 GWF2RIP3AR

The subroutine GWF2RIP3AR will read from the RIP-ET file designated in the NAME file, allocates space for RIP-ET array, writes to the list file, and performs the following steps:

Step 1: Allocate scalar variables, which make it possible for multiple grids to be defined.

Step 2: Identifies the package as the Riparian Package, and initializes the number of active riparian cells (NRIPCL) to zero.

Step 3: Read in the maximum number of riparian cells (MAXRIP) allowed over all stress periods. Note that NRIPCL, which is updated every stress period, is always ≤MAXRIP. Read in maximum number of polygons in a cell (MAXPOLY). Read in the cell-by-cell flow flag or unit number (IRIPCB), and the flag or unit number for plant functional subgroup ET rate storage (IRIPCB1).

Step 4: Read in the maximum number of plant functional subgroups (MAXTS), and maximum number of segments used to interpolate the transpiration canopy flux function (MXSEG).

Step 5: Allocate space for the real variables: saturated extinction depth (Sxd), the active root depth (Ard), the maximum ET canopy flux (Rmax), the ET canopy flux at the saturation extinction depth (Rsxd), the active root depth segment factor (fdh), the maximum ET flux segment factor (fdR), the land surface elevation for a polygon within a cell (HSURF), the PFSG fractional coverage in a cell for a polygon (CovPFSG), the total ET rate for a cell (RIPET).

Step 6: Allocate space for integer variables: the number of active segments for a PFSG (NuSeg), the $(k, i, j)$ location of the cell (CLoc), the number of polygons in a cell (NPOLY).

Step 7: Allocate space for double precision variables: the HCOF portion of ET rate (C1), the RHS portion of ET rate (C2), and the ET rate for the PFSG within a polygon for a cell (ETR).

Step 8: Allocate space for character variables: the PFSG names (RIPNM) arrays.

Step 9: For each plant functional subgroup, read the name, saturated extinction depth (measured from land surface), active root depth, maximum ET flux, ET flux at the saturation extinction depth, and the number of segments used to interpolate the ET flux curve.

Step 10: For each plant functional subgroup, read the active root depth segment factors and ET flux segment.

Step 11: Print out plant functional subgroup name, saturated extinction depth (measured from land surface), active root depth, maximum ET flux, ET flux at saturation extinction depth.

Step 12: Print out plant functional subgroup name and active root depth and ET flux segments.

Step 13: Save pointers to data.

Step 14: Return.

The variables used in GWF2RIP3AR are listed in table 7.

**Table 7.** List of variables in module GWF2RIP3AR.

| Variable | Scope | Definition |
|---|---|---|
| LINE | Module | Content of one line read from the input file |
| IFREFM | Global | Flag indicating if variables are to be read in free format |
| IN | Package | Unit number of input file |
| IOUT | Global | Unit number of LIST output file |
| LLOC | Module | Pointer used to keep track of position in LINE |
| ISTART | Module | Starting position of parsed word |
| ISTOP | Module | Ending position of parsed word |
| N | Module | Dummy integer variable |
| R | Module | Dummy real variable |
| NRIPCL | Package | Active number of riparian cells in a stress period |
| MAXRIP | Package | Maximum number of riparian cells |
| MAXPOLY | Package | Maximum number of polygons in a cell |
| IRIPCB | Package | Cell-by-cell flag and unit number |
| IRIPCB1 | Package | Save flag and unit number |
| MAXTS | Package | Maximum number of plant functional subgroups |
| MXSEG | Package | Maximum number of segments over all subgroups |
| Sxd | Package | Saturated extinction depth |
| Ard | Package | Active root depth |
| Rmax | Package | Maximum ET canopy flux |
| Rsxd | Package | ET canopy flux at the saturation extinction depth |
| fdh | Package | Active root depth segment factor |
| fdR | Package | Maximum ET flux segment factor |
| HSURF | Package | Land-surface elevation for a polygon within a cell |
| CovPFSG | Package | PFSG fractional coverage in a cell for a polygon |
| RIPET | Package | Total ET rate for a cell |
| NuSeg | Package | Number of active segments for a PFSG |
| LCov | Package | The $(k,i,j)$ location of the cell |
| NPOLY | Package | Number of polygons in a cell |
| C1 | Package | HCOF portion of ET rate |
| C2 | Package | RHS portion of ET rate |
| ETR | Package | ET rate for the PFSG within a polygon for a cell |
| RIPNM | Package | PFSG names |

## 15.5 Listing for GWF2RIP3AR

```
|     ******************************************************************
|     Subroutine GWF2RIP3AR(IN,IGRID)
|     ******************************************************************
|     This subroutine allocates storage for all RIP-ET data, and reads in
|     dimension variables that remain constant over the simulation
|     The follow variables are used:
|       IN is the unit number of input file.
|       IGRID is the grid number for LGR.
|       IOUT is the unit number of the LIST output file.
|       IFREFM is a flag indicating free format for data input.
|         MAXRIP is the maximum number of riparian cells over all stress
|         periods.
|       MAXPOLY is the maximum number of polygons in a cell.
|       IRIPCB is a cell-by-cell printing flag and unit number,
|       IRIPCB1 is a flag to save cell values:location, surface elevation.
|         and ET rate for each plant functional subgroup.
|       NRIPCL is the number of active riparian cells in a stress period.
|       MAXTS is the total number of plant functional subgroups for
|         all cells.
|       MXSEG is the maximum number of segments for interpolation of ET
|         flux rate as a function of hydraulic head.
|       NuSEG is the number of active segments for a plant functional
|         subgroup.
|       Sxd is the saturation extinction depth with respect to land
|         surface.
|       Ard is the active root depth.
|       Rmax is the maximum ET canopy flux rate.
|       Rsxd is the ET canopy flux rate at the saturation extinction
|         depth.
|       CLoc has the (k,i,j) location of the cell for a stress period
|       NPoly is the number of polygons in a cell for a stress period.
|           fdh is the active root depth segment factor.
|           fdR is the maximum ET flux segment factor.
|       HSurf is the land surface elevation for a polygon within a cell
|       CovPFSG has the PFSG fractional coverage in a cell for a polygon for
|         a stress period.
|       RIPET is the total ET rate for a cell.
|       C1 is the HCOF portion of ET rate.
|       C2 is the RHS portion of ET rate.
|       RIPNM is the plant functional subgroup names, may be up to
|         24 characters in length.
|       LINE is the content of one line read from the input file.
|       LLOC is a pointer used to keep track of position in LINE.
|       ISTART is the starting position of parsed word.
|       ISTOP is the ending position of parsed word.
|       N is a dummy integer variable.
|       R is a dummy real variable.
|       Nfdh is the characters 'fdh'.
|       NfdR is the characters 'fdR'.
|       ITS and ISEG are counters.
|     ******************************************************************
|
|     Specifications
|     ******************************************************************
      Use GLOBAL,       Only: IOUT,IFREFM
        Use GWFRIPMODULE, Only: MAXRIP,MAXPOLY,IRIPCB,IRIPCB1,NRIPCL,MAXTS,     &
                                MXSEG,NuSeg,Sxd,Ard,Rmax,Rsxd,CLoc,NPoly,       &
                                fdh,fdr,HSurf,CovPFSG,RIPET,C1,C2,RIPNM
          Integer            :: IN,IGRID,LLOC,ISTART,ISTOP,N,ITS,ISEG
          Real               :: R
          Character(3)       :: Nfdh='fdh',NfdR='dfR'
      Character(80)      :: Line
|     ******************************************************************
|
|1----Allocate scalar variables, which make it possible for multiple grids
|     to be defined.
|
```

```
|
|2----Identify package and initialize NRIPCL
|
        Write(IOUT,100) IN
100    Format(1x,/1x,'RIP-ET -- RIPARIAN PACKAGE, VERSION 3,', &
                      ' 6/21/2010',' INPUT READ FROM UNIT',I3)
|
        NRIPCL=0
|
|3----Read 1) maximum number of riparian cell, 2) the maximum
|     number of polygons per cell 3) a unit or flag for
|     printing or storing ET terms, and 4) an option value,if
|     present,that allows storage in memory of the total ET Rate
|        for each cell.
|
      Read(IN,'(A)') LINE
      If (IFREFM == 0) Then
             Read(LINE,'(4I10)') MAXRIP,MAXPOLY,IRIPCB,IRIPCB1
             LLOC=41
        Else
             LLOC=1
             CALL URWORD(LINE,LLOC,ISTART,ISTOP,2,MAXRIP,R,IOUT,IN)
             CALL URWORD(LINE,LLOC,ISTART,ISTOP,2,MAXPOLY,R,IOUT,IN)
             CALL URWORD(LINE,LLOC,ISTART,ISTOP,2,IRIPCB,R,IOUT,IN)
             CALL URWORD(LINE,LLOC,ISTART,ISTOP,2,IRIPCB1,R,IOUT,IN)

        Endif
        Write(IOUT, 110) MAXRIP
110    Format(1x,'MAXIMUM OF',I5,' RIPARIAN CELLS')
        Write(IOUT,120) MAXPOLY
120    Format(1x,'MAXIMUM OF',I5,' POLYGONS PER CELL')
          If(IRIPCB < 0) then
                write(IOUT, 130)
130     Format(1x,'THE ET RATE FOR EACH PLANT FUNCTIONAL SUBGROUP WILL ', &
                   'BE WRITTEN TO THE LIST FILE WHEN ICBCFL IS NOT 0')
          Elseif (IRIPCB > 0) then
                write(IOUT,140)IRIPCB
140     Format(1x,'TOTAL ET RATE FOR EACH CELL WILL BE SAVED ON UNIT',I3)
          Endif
          If(IRIPCB1>0) Then
                Write(IOUT,150) IRIPCB1
150     Format(1x,'CELL LOCATION, SURFACE ELEVATION, AND PLANT ',        &
                   'FUNCTIONAL SUBGROUP ET RATES WILL BE SAVED ',        &
                               'ON UNIT',I3)
          End if
|
|4----Read and print maximum number of plant functional subgroups (MAXTS),
|     and the maximum of interpolation segments for the ET flux
|     function (MXSEG).
|
      If (IFREFM == 0) Then
        Read(IN,'(2I10)') MAXTS,MXSEG
      Else
        Read(IN,*) MAXTS,MXSEG
      Endif
        write(IOUT,160) MAXTS,MXSEG
160    Format(1x,'MAXIMUM NUMBERS OF PLANT FUNCTIONAL SUBGROUPS AND SEGMENTS '/, &
            1x,'PLANT FUNCTIONAL SUBGROUPS  =',I3,/, &
                1x,'MAXIMUM CURVE SEGMENTS      =',I3)
|
|5----Allocate REAL variables: Sxd, Ard, Rmax, Rsxd, fdh, fdR, HSurf, RIPET
|     CovPFSG.
|
      Allocate(Sxd(MAXTS),Ard(MAXTS),Rmax(MAXTS),Rsxd(MAXTS),RIPET(MAXRIP))
      Allocate(fdh(MAXTS,MXSEG),fdR(MAXTS,MXSEG),HSurf(MAXRIP,MAXPOLY))
      Allocate(CovPFSG(MAXRIP,MAXPOLY,MAXTS))
|
|6----Allocate INTEGER variables: NuSeg,Cloc,NPOLY
|
        Allocate(NuSeg(MAXTS),NPoly(MAXRIP) )
        Allocate(Cloc(MAXRIP,3))
```

```
!
!7----Allocate DOUBLE PRECISION variables: C1,C2
!
      Allocate(C1(MAXRIP,MAXPOLY,MAXTS),C2(MAXRIP,MAXPOLY,MAXTS))
!
!8----Allocate CHARACTER variables:
!
          Allocate(RIPNM(MAXTS))
!
!9----For each plant functional subgroup, read the name, saturated extinction
!     depth (measured from land surface), active root depth, maximum ET flux,
!     ET flux at the saturation extinction depth, and the number of
!     segments used to interpolate the ET flux curve.
!
      Do ITS=1, MAXTS
              If(IFREFM == 0) then
                  Read(In,'(A,4F10.0,I10)') RIPNM(ITS),Sxd(ITS),Ard(ITS),    &
                             Rmax(ITS),Rsxd(ITS), NuSEG(ITS)
              Else
          Read(IN,*)RIPNM(ITS),Sxd(ITS),Ard(ITS),Rmax(ITS),Rsxd(ITS),        &
                             NuSEG(ITS)
          End if
!
!10---For each plant functional subgroup, read the active root
!     depth segment factors and ET flux segment factors.
!
              If(IFREFM == 0) then
                  Read(In,'(10F10.4)') (fdh(ITS,ISEG), ISEG=1,NuSEG(ITS))
          Read(In,'(10F10.4)') (fdR(ITS,ISEG), ISEG=1,NuSEG(ITS))
              Else
            Read(IN,*) (fdh(ITS,ISEG), ISEG=1,NuSeg(ITS))
            Read(IN,*) (fdR(ITS,ISEG), ISEG=1,NuSeg(ITS))
          End if
          End do
!
!11----Print out plant functional subgroup name, saturated extinction depth
!      (measured from land surface),active root depth, maximum ET flux,
!      ET flux at saturation extinction depth.
!
          Write(IOUT,170)
170   Format(/,'              RIPARIAN INFORMATION'//,    &
          '              NAME          SATURATION    ACTIVE      ',  &
                        'MAXIMUM    ET FLUX AT'/,                                &
                    '                                  EXTINCTION    ROOT       ',  &
                    'ET      SATURATION'/,                                          &
                    '                                  DEPTH      DEPTH       ',  &
                    'FLUX    EXTINCTION DEPTH')
!
          Do ITS=1, MAXTS
          Write(IOUT,180)RIPNM(ITS),Sxd(ITS),Ard(ITS),Rmax(ITS),Rsxd(ITS)
180       Format(1x,A,T27,F10.4,T38,F10.4,T52,E11.4,T65,E11.4)
      End do
!
!12---Print out plant functional subgroup name and active root depth and ET
!     flux segments.
!
          Write(IOUT,190)
190   Format(///,'              SEGMENT INFORMATION'//,                  &
                    '              NAME                          SEGMENTS ')
!
          Do ITS=1, MAXTS
                  Write(IOUT,200)RIPNM(ITS),Nfdh,(fdh(ITS,ISEG), ISEG=1,NuSEG(ITS))
200       Format(/,1x,A,T27,A,10F10.4)
          Write(IOUT,210) NfdR,(fdR(ITS,ISEG), ISEG=1,NuSEG(ITS))
210       Format(1x,T27,A,10F10.4)
      End Do
!
!13---Save pointers to data
!
      Call SGWF2RIP3PSV(IGRID)
!
!14---Return
!
          Return
      End Subroutine GWF2RIP3AR
```

## 15.6 GWF2RIP3RP

Subroutine GWF2RIP3FM performs the following steps:

Step 1: Set pointers for current grid.

Step 2: Read in ITMP (number of riparian cells or flag to reuse data).

Step 3: Test ITMP.

Step 3a: If ITMP < 0, reuse data from last stress period.

Step 3b: If ITMP ≥ 0, it is the number of riparian cells for the stress period.

Step 4: Check that number of riparian cell for the stress period does not exceed MAXRIP.

Step 5: Printout number of active riparian cells for the stress period.

Step 6: If there are no riparian cells this stress period, return.

Step 7: Initialize NPOLY, CLOC, HSURF, and COVPFSG.

Step 8: For each riparian cell, read a cell number and the number of polygons in the cell.

Step 9: For each polygon, read surface elevation, and then fractional coverage for each plant functional subgroup. A polygon is limited to 24 PFSG unless the format is changed.

Step 10: Print output heading.

Step 11: Printout polygon data cell by cell.

Step 12: Return.

The variables used in GWF2RIP3RP are listed in table 8.

**Table 8.** List of variables in module GWF2RIP3RP.

| Variable | Scope | Definition |
|---|---|---|
| ITMP | | Number of riparian cells or flag to reuse data |
| IFREFM | Global | Flag indicating if variables are to be read in free format |
| IN | Package | Unit number of input file |
| IOUT | Global | Unit number of LIST output file |
| MAXTS | Package | Maximum number of plant functional subgroups |
| MXSEG | Package | Maximum number of segments over all subgroups |
| MAXRIP | Package | Maximum number of riparian cells |
| MAXPOLY | Package | Maximum number of polygons in cells |
| NRIPCL | Package | Active number of riparian cells in a stress period |
| CLoc | Package | The $(k,i,j)$ location of the riparian cell |
| HSURF | Package | Land surface elevation for a polygon within a cell |
| NPOLY | Package | Number of polygons in a cell |
| CovPFSG | Package | PFSG fractional coverage in a cell for a polygon |
| K | Module | Layer |
| J | Module | Column |
| I | Module | Row |
| HSURF | Module | Land-surface elevation |
| II | Module | Counter |
| JJ | Module | Counter |
| NN | Module | Counter |

## 15.7 Listing for GWF2RIP3RP

```
|      ******************************************************************
|      Subroutine GWF2RIP3RP(IN,IGRID)
|      ******************************************************************
|      This subroutine reads by stress period, the number of riparian cells,the
|      number of polygons for the cell,the location of the riparian cells
|      (layer, row, column), polygon number,the land surface elevation for the
|      polygon,and the percentage coverage of each plant functional subgroup in
|      each polygon.
|         IN      is the unit number of input file.
|         IGRID   is the grid number for LGR.
|         IOUT    is the unit number of the LIST output file.
|         IFREFM  is a flag indicating free format for data input.
|                 stress period.
|          MAXRIP  is the maximum number of riparian cells over all stress
|                  periods.
|         MAXPOLY is the maximum number of polygons in a cell.
|         MAXTS   is the total number of plant functional subgroups for
|                 all cells.
|         NRIPCL  is the number of active riparian cells in a stress period.
|         CLoc    has the (k,i,j) location of the cell for a stress period
|         NPoly   is the number of polygons in a cell for a stress period..
|         HSurf   is the land surface elevation for a polygon within a cell
|         CovPFSG has the PFSG fractional coverage in a cell for a polygon for
|                 a stress period.
|         RIPNM   is the plant functional subgroup names, may be up to
|                 24 characters in length.
|           ITMP     is number of riparian cells or flag to repeat data from previous
|         NC,NP,and I are counters.
|      ******************************************************************
|
|      Specifications
|      ******************************************************************
|      Use Global,       Only: IOUT,IFREFM
|         Use GWFRIPMODULE,  Only: MAXRIP,MAXPOLY,MAXTS,NRIPCL,CLoc,NPoly,HSurf,  &
|                               CovPFSG,RIPNM
|           Integer:: IN,IGRID,ITMP
|           Integer:: NC,NP,I
|      ******************************************************************
|
|1----Set pointers for current grid
|
|      Call SGWF2RIP3PNT(IGRID)
|
|2----Read in ITMP (number of riparian cells or flag to reuse data).
|
|           If(IFREFM == 0) then
|                Read(IN,'(I10)') ITMP
|           Else
|                Read(IN,*) ITMP
|           Endif
|
|3----Test ITMP.
|
|           If(ITMP < 0) Then
|
|3a---If ITMP<0, reuse data from last stress period.
|
```

```
                        write(IOUT,100)
100                     Format(1x,/1x,'REUSING RIPARIAN CELL DATA FROM LAST STRESS PERIOD')
                        Return
              Else
|
|3b---If ITMP>=0, it is the number of riparian cells.
|
          NRIPCL=ITMP
             End if
|
|4----If NRIPCL>MAXRIP then STOP.
|
          If(NRIPCL > MAXRIP) Then
                    write(IOUT,110) NRIPCL,MAXRIP
110                 Format(1x,/,1x,'NRIPCL(',I4,') IS GREATER THAN MAXRIP(',I4,')')
                    Stop
          Endif
|
|5----Printout number of active riparian cells for the stress period.
|
          write(IOUT,120) NRIPCL
120       Format(1x,//,1x,I5,' RIPARIAN CELLS.')
|
|6----If there are no riparian cells this stress period, return.
|
      If(NRIPCL == 0) Return
|
|7----Initialize NPOLY, CLOC, HSURF, and COVPFSG
|
          NPOLY   = 0
          CLOC    = 0
          HSURF   = 0.0
          COVPFSG = 0.0
|
|8----For each riparian cell, read a cell number and the number of polygons
|      in the cell.
|
          Do NC=1, NRIPCL
        If(IFREFM == 0) then
          Read(IN,'(4I10)') (CLOC(NC,I),I=1,3),NPOLY(NC)
        Else
              Read(IN,*) (CLOC(NC,I),I=1,3),NPOLY(NC)
            End If
|
|9----For each polygon, read surface elevation, and then fractional
|      coverage for each plant functional subgroup. A polygon is limited
|      to 24 PFSG unless the format is changed.
|
      Do NP =1,NPOLY(NC)
        If(IFREFM == 0) then
                Read(IN,'(F10.2,25F10.5)') HSURF(NC,NP),(COVPFSG(NC,NP,I),I=1,MAXTS)
              Else
                Read(IN,*) HSURF(NC,NP),(COVPFSG(NC,NP,I),I=1,MAXTS)
              Endif
          End Do   !End of polygon loop
    End Do   !End of cell loop
|
|10---Print output heading.
```

## 15.8 GWF2RIP3FM

Subroutine GWF2RIP3FM performs the following steps:

Step 1: Set pointers to current grid.

Step 2: Check to see if there are riparian cells for this stress period, if not, return.

Step 3: Begin to process each cell in the riparian cell list.

Step 4: Get row, column and layer of riparian cell from CLoc array.

Step 5: Check to see if cell is external; if so, skip it and send a message that it is in an inactive cell (IBOUND=0).

Step 6: Set head for cell (HH=HNEW).

Step 7: Begin to process each polygon in a cell.

Step 8: Begin to process each plant functional subgroup in a cell.

Step 9: Initialize C1 and C2, and determine which plant functional subgroups are active within a cell (RIP≠0).

Step 10: Determine cell non-zero fractional coverage for a plant functional subgroup, calculate Hsxd and Hxd; and initialize the extinction depth segment end points, HK(1) to Hxd, and RK(1) to zero.

Step 11: Check to see if HH is beyond the ends of the ET canopy flux curve; if HH <= Hxd, set both C1 and C2 to zero; or if HH >= Hsxd, set C1=0.0 and C2=-Rsxd*fCov*DELC(IR)*DELR(IC); in either case, cycle to the next plant functional subgroup.

Step 12: Loop through the ET canopy flux function vertices.

Step 13: Calculate the HKs and RKs as needed.

Step 14: Check to see if HK(KS)<HH≤HK(KS+1).

Step 15: When it is, calculate C1 and C2 using the functions HCOFtrm and RHStrm that are adjusted with fCov, DELR and DELC, i.e.,

C1=fCov*HCOFtrm*DELR*DELC

C2=fCov*RHStrm*DELR*DELC.

Step 16: Add C1 to HCOF, and subtract C2 from RHS.

Step 17: Return

The variables used in GWF2RIP3FM are listed in table 9.

**Table 9.**    List of variables in module `GWF2RIP3FM`.

| Variable | Scope | Definition |
|---|---|---|
| NLAY | Global | Number of layers in the grid |
| NCOL | Global | Number of columns in the grid |
| NROW | Global | Number of rows in the grid |
| HNEW | Global | Most recent estimate of head in a cell |
| HCOF | Global | Coefficient of head in the finite-difference equations |
| RHS | Global | Right-hand side of the finite-difference equations |
| DELR | Global | Cell dimension in row direction |
| DELC | Global | Cell dimension in column direction |
| IBOUND | Global | Status of cell: $<0$, constant head; $=0$, inactive; $>0$, active |
| NRIPCL | Package | Active number of riparian cells in a stress period |
| MAXRIP | Package | Maximum number of riparian cells |
| MAXTS | Package | Maximum number of plant functional subgroups |
| MXSEG | Package | Maximum number of segments over all subgroups |
| CLoc | Package | The $(k,i,j)$ location of the riparian cell |
| HSURF | Package | Land-surface elevation for a polygon within a cell |
| NPOLY | Package | Number of polygons in a cell |
| CovPFSG | Package | PFSG fractional coverage in a cell for a polygon |
| HH | Module | Hydraulic head |
| HK | Package | Depth segment vertex component |
| RK | Package | ET Rate segment vertex component |
| Sxd | Package | Saturation extinction depth |
| Ard | Package | Active root depth |
| Rmax | Package | Maximum ET canopy flux |
| Rsxd | Package | ET canopy flux at saturation extinction depth |
| NuSeg | Package | Number of segments for a plant functional subgroup ET curve |
| fdh | Package | Active root depth segment factor |
| fdR | Package | Maximum ET flux segment factor |
| Hxd | Module | Extinction depth elevation |
| Hsxd | Module | Saturation extinction depth elevation |
| C1 | Package | HCOF portion of ET rate |
| C2 | Package | RHS portion of ET rate |
| LC | Module | Counter |
| IL | Module | Layer number |
| IR | Module | Row number |
| IC | Module | Column number |
| KS | Module | Counter |
| NTS | Module | Counter |
| Loc1 | Module | Counter for fractional coverage values in a cell. |
| Loc2 | Module | Counter for land surface elevations in a cell, used only if IHSURF $>0$. |
| HCOFtrm | Package | Function for HCOF portion of ET rate |
| RHStrm | Package | Function for RHS portion of ET rate |

## 15.9 Listing for GWF2RIP3FM

```
| *************************************************************************
  Subroutine GWF2RIP3FM(IGRID)
| *************************************************************************
| Add riparian evapotranspiration to RHS and HCOF. The ET flux
| rate is estimated using a segmented interpolation structure.
|    IGRID is the grid number for LGR.
|    NCOL is the number of columns in the grid.
|    NROW is the number of rows in the grid.
|    NLAY is the umber of layers in the grid.
|    DELR is the cell dimension in row direction.
|    DELC is the cell dimension in the column direction.
|    IBOUND is status of cell: <0, constant head;=0, inactive;>0, active.
|    HNEW is the most recent estimate of head in a cell.
|    RHS is the right-hand side of the finite-difference equations.
|    HCOF is the coefficient of head in the finite-difference equations.
|    MAXTS is the total number of plant functional subgroups for
|       all cells.
|         MXSEG is the maximum number of segments for interpolation of ET
|       flux as a function of hydraulic head.
|    NRIPCL is the number of active riparian cells in a stress period.
| Sxd is the saturation extinction depth with respect to land
|       surface.
| Ard is the active root depth.
| Rmax is the maximum ET canopy flux.
| Rsxd is the ET canopy flux at the saturation extinction
|       depth.
|         fdh is the active root depth segment factor.
|         fdR is the maximum ET flux segment factor.
| NuSEG is the number of active segments for a plant functional
|       subgroup.
| CLoc has the (k,i,j) location of the cell for a stress period
| NPoly is the number of polygons in a cell for a stress period.
| HSurf is the land surface elevation for a polygon within a cell
| CovPFSG has the PFSG fractional coverage in a cell for a polygon for
|       a stress period.
| C1 is the HCOF portion of ET rate.
| C2 is the RHS portion of ET rate.
| IL is the cell layer.
| IR is the cell row.
| IC is the cell column.
| HH is the hydraulic head in the cell.
| HK(N) is the head value at Nth vertex of the Ard segments.
| RK(N) is the ET canopy flux at Nth vertex of the R segments.
| HCOFtrm is a function to calculate HCOF part of ET canopy flux.
| RHStrm is a function to calculate RHS part of ET canopy flux.
| fCov is the fractional coverage in a cell by a plant functional
|       subgroup.
|         Hxd is the extinction depth elevation.
| Hsxd is the saturated extinction depth elevation.
| LC,LP,KS,and NTS are counters.
```

```
|      *************************************************************************
|
|      Specifications:
|      *************************************************************************
|      Use GLOBAL,          Only:NCOL,NROW,NLAY,DELR,DELC,IBOUND,HNEW,RHS,HCOF
|      USE GWFRIPMODULE, Only:MAXTS,MXSEG,NRIPCL,Sxd,Ard,Rmax,Rsxd,fdH,fdR,       &
|                             NuSeg,CLoc,NPoly,HSurf,CovPFSG,C1,C2
|      Integer::IGRID
|         Integer::IL,IR,IC
|         Integer::LC,LP,KS,NTS
|         Double Precision::HH
|         Double Precision, Dimension(MXSEG+1)::HK,RK
|         Double Precision::HCOFtrm,RHStrm
|         Real::fCov,Hxd,Hsxd
|      *************************************************************************
|
|1----Set pointers to current grid.
|
|      Call SGWF2RIP3PNT(IGRID)
|
|2----If NRIPCL<=0, there are no riparian cells, return.
|
|         If (NRIPCL<=0) return
|
|3----Process each cell in the riparian cell list.
|
|      Cell_Loop: Do LC=1,NRIPCL
|
|4----Get column, row and layer from riparian cell array.
|
|             IL=CLoc(LC,1)
|             IR=CLoc(LC,2)
|             IC=CLoc(LC,3)
|
|5----If cell is external skip it and send a message to screen that cell
|     is inactive.
|
|             If(IBOUND(IC,IR,IL) == 0) Then
|                 Write(*,'(A)') '**** Riparian cell is inactive ****'
|           Cycle Cell_Loop
|         End if
|
|6----Set head for cell.
|
|             HH=HNEW(IC,IR,IL)
|
|7----Process each polygon in a cell
|
|      Poly_Loop: Do LP=1,NPoly(LC)
|
|8----Process each plant functional subgroup
|
|       TS_Loop:Do NTS=1,MAXTS
|
|9----Initialize C1 and C2
|
|                 C1(LC,LP,NTS)=0.0
|                 C2(LC,LP,NTS)=0.0
|
|10---Determine cell non-zero fractional coverage for a plant functional
|     subgroup, calculate Hsxd and Hxd; and initialize the extinction
|     depth segment end points, HK(1) to Hxd, and RK(1) to zero.
|
|                 fCov=CovPFSG(LC,LP,NTS)
|           Hsxd=HSURF(LC,LP)-Sxd(NTS)
|           Hxd=Hsxd-Ard(NTS)
|           HK(1)=Hxd
|           RK(1)=0.0
```

```
|
|11---Check to see if HH is beyond the ends of the ET canopy flux curve;
|    If HH <= Hxd, set both C1 and C2 to zero; or if HH >= Hsxd,
|    set C1=0.0 and C2=-Rsxd*fCov*DELC(IR)*DELR(IC);
|    in either case, cycle to the next plant functional subgroup.
|
                    If(HH > Hsxd ) Then
                         C1(LC,LP,NTS)=0.0
                         C2(LC,LP,NTS)=-Rsxd(NTS)*fCov*DELR(IC)*DELC(IR)
                         RHS(IC,IR,IL)=RHS(IC,IR,IL)-C2(LC,LP,NTS)
                         cycle TS_Loop
                    Else if(HH <= Hxd) then
                         C1(LC,LP,NTS)=0.0
                         C2(LC,LP,NTS)=0.0
                         cycle TS_Loop
                    End if
|
|12---Loop through the ET canopy flux function vertices.
|
            Seg_Loop:DO KS=1, NuSeg(NTS)
|
|13---Calculate HKs and RKs as needed.
|
                    HK(KS+1)=Hk(KS)+fdh(NTS,KS)*Ard(NTS)
                    RK(KS+1)=RK(KS)+fdR(NTS,KS)*Rmax(NTS)
|
|14----Check to see if HK(KS)<HH<=HK(KS+1)
|
                    IF(HH > HK(KS) .and. HH <= HK(KS+1)) Then
|
|15----When it is, calculate C1 and C2 using the functions HCOFtrm and
|    RHStrm that are adjusted with fCov, DELR, and DELC.
|
                    C1(LC,LP,NTS)=-fCov*HCOFtrm(HK(KS),HK(KS+1),    &
                                         RK(KS),RK(KS+1))* DELR(IC)*DELC(IR)
                    C2(LC,LP,NTS)=-fCov*RHStrm(HK(KS),HK(KS+1),     &
                                         RK(KS),RK(KS+1))*DELR(IC)*DELC(IR)
|
|16----Add C1 to HCOF and subtract C2 from RHS.
|
                    HCOF(IC,IR,IL)=HCOF(IC,IR,IL)+C1(LC,LP,NTS)
                    RHS(IC,IR,IL)=RHS(IC,IR,IL)-C2(LC,LP,NTS)
                    EXIT Seg_Loop
                    End if
                End do Seg_Loop
            End do TS_Loop
        End Do Poly_Loop
        End do Cell_Loop
|
|17-----Return
|
        Return
|
        End Subroutine GWF2RIP3FM
```

## 15.10 GWF2RIP3BD

Subroutine GWF2RIP3BD performs the following steps:

Step 1: Set pointers for current grid.

Step 2: Initialize the rate accumulator (RATOUT).

Step 3: Set cell-by-cell budget save flag (IBD) and clear buffer.

Step 4: If cell-by-cell flows will be saved as a list, write header.

Step 5: Initialize BUFF.

Step 6: Loop through each riparian cell (if NRIPCL > 0).

Step 7: Get row, column, layer.

Step 8: Set HH=HNEW and initialize RATE to zero.

Step 9: Loop through the polygons.

Step 10: Loop through plant functional groups, calculate ETR, and aggregate ETR to RAT.

Step 11: Load RATE to budget accumulator and to BUFF and RIPET.

Step 12: If saving cell-by-cell flows to list, write flows.

Step 13: Calculate the total ET for each PFSG.

Step 14: Print locations, polygon number, HSURF, ET rate by plant functional groups, and total ET rate per cell, if requested.

Step 15: Write out ET rates to list file.

Step 16: Check to see if cell-by-cell flows are to be saved as three-dimensional array. The values are stored as a binary file.

Step 17: Check to see if cell location, polygon, head, surface elevation, and plant functional subgroup ET rates are to be saved as a formatted file.

Step 18: Move total riparian loss into VBVL for printing in BAS1OT.

Step 19: Add riparian ET (riparian ET rate times time's step length) to VBVL.

Step 20: Move budget labels to VBNM for print by module BAS1OT.

Step 21: Increment Budget term counter.

Step 22: Return.

The variables used in GWF2RIP3BD are listed in table 10.

**Table 10.**   List of variables in module `GWF2RIP3BD`.

| Variable | Scope | Definition |
|---|---|---|
| NLAY | Global | Number of layers in the grid |
| NCOL | Global | Number of columns in the grid |
| NROW | Global | Number of rows in the grid |
| HNEW | Global | Most recent estimate of head in a cell |
| MSUM | Global | Counter for budget entries and labels in VBVL and VBNM |
| KSTP | Global | Time step counter |
| KPER | Global | Stress period counter |
| RATOUT | Module | Accumulator for total flow out of flow field to riparian ET |
| IBOUND | Global | Status of cell: <0, constant head; =0, inactive; >0, active |
| VBNM | Global | Labels for entries in the volumetric budgets |
| VBVL | Global | Entries for volumetric budget. |
| DELT | Global | Length of current time step |
| RATE | Module | ET in cell aggregated over plant functional subgroups |
| ZERO | Module | The number 0 |
| HH | Module | Hydraulic head |
| TEXT | Module | Label to be printed or recorded with array data. |
| BUFF | Global | Buffer used to accumulate information before printing it |
| IOUT | Global | Primary unit number for all printed output |
| IBD | Module | Flag for recording cell-by-cell flows |
| ICBCFL | Global | Flag for recording cell-by-cell flows |
| IRIPCB | Package | Flag for recording cell-by-cell flows |
| IRIPCB1 | Package | Flag for list file printing |
| NRIPCL | Package | Active number of riparian cells in a stress period |
| MAXRIP | Package | Maximum number of riparian cells |
| MAXTS | Package | Maximum number of plant functional subgroups |
| IBDLBL | Module | A flag that indicate the first pass through a loop |
| CLoc | Package | The $(k,i,j)$ location of the riparian cell |
| HSURF | Package | Land surface elevation for a polygon within a cell |
| NPOLY | Package | Number of polygons in a cell |
| ETR | Package | ET rate for the plant functional subgroup within a polygon for a cell |
| RIPET | Package | Total ET rate for a cell |
| C1 | Package | HCOF portion of ET rate |
| C2 | Package | RHS portion of ET rate |
| IL | Module | Layer number |
| IC | Module | Row number |
| IR | Module | Column number |
| NTS | Module | Counters |
| II | Module | Counters |

## 15.10 Listing for GWF2RIP3BD

```
|    ***************************************************************************
     Subroutine GWF2RIP3BD(KSTP,KPER,IGRID)
|    ***************************************************************************
|    Calculates volumetric budgets for riparian cells.
|      KSTP is a time step counter.
|      KPER is a stress period counter.
|      IGRID is the grid number for LGR.
|      IOUT is the unit number of the LIST output file.
|      NCOL is the number of columns in the grid.
|      NROW is the number of rows in the grid.
|      NLAY is the umber of layers in the grid.
|      IBOUND is status of cell: <0, constant head;=0, inactive;>0, active.
|      HNEW is the most recent estimate of head in a cell
|      BUFF is a three dimensional array containing the RATE for each
|           cell.
|      MSUM is the Counter for budget entries and labels in VBVL and VBNM
|      VBVL are the entries for volumetric budget.
|      VBNM are the abels for entries in the volumetric budgets
|      ICBCFL is the flag for recording cell-by-cell flows
|      DELT is the length of current time step
|      MAXRIP is the maximum number of riparian cells over all stress
|         periods.
|      MAXPOLY is the maximum number of polygons in a cell.
|      MAXTS is the otal number of plant functional subgroups for
|         all cells
|      NRIPCL is the number of active riparian cells in a stress period
|      IRIPCB is a cell-by-cell save flag and unit number
|      IRIPCB1 is a flag to save cell values:location, surface elevation
|         and ET rate for each plant functional subgroup.
|      C1 is the HCOF portion of ET rate.
|      C2 is the RHS portion of ET rate.
|      CLoc has the (k,i,j) location of the cell for a stress period
|      NPoly is the number of polygons in a cell for a stress period.
|      HSurf is the land surface elevation for a polygon within a cell
|      CovPFSG has the PFSG fractional coverage in a cell for a polygon for
|      RIPNM is the plant functional subgroup names, may be up to
|         24 characters in length.
|      RIPET is the total ET rate for a cell.
|      TEXT is a 16 character string, '      RIPARIAN ET' used in the
|         budget.
|      IL is the cell layer.
|      IR is the cell row.
|      IC is the cell column.
|      IBD is a print or save flag.
|      Zero is the number 0.
|      RATE is plant functional subgroup aggregated ET rate for a cell,
|              and is written to the RIP array if CBC is indicated under
|         options.
|      Sum is an accumulator.
|      TotPFSG is the total ET for each PFSG.
|      RATOUT is the Accumulator for total flow out of flow field
|         to riparian ET.
|      HH is the hydraulic head in a cell
|         ETR is the ET rate for the plant functional subgroup within a
|             polygon for a cell.
|      LC,LP,NTS and II are counters
```

```
|     ***********************************************************************
|
|     Specifications:
|     ***********************************************************************
      Use GLOBAL,          Only: IOUT,NCOL,NROW,NLAY,IBOUND,HNEW,BUFF
      Use GWFBASMODULE,  Only: MSUM,VBVL,VBNM,ICBCFL,DELT,PERTIM,TOTIM
      USE GWFRIPMODULE,  Only: MAXRIP,MAXPOLY,MAXTS,NRIPCL,IRIPCB,IRIPCB1,      &
                               C1,C2,CLoc,NPoly,HSurf,CovPFSG,RIPNM,RIPET
      Character(16)                    :: TEXT,RIPAUX=' '
         Integer                       :: KSTP,KPER,IGRID,NAUX=0
         Integer                       :: IL,IC,IR,LC,LP,NTS,II,IBD,NP
         Real                          :: Zero,RATE,Sum,AUXVAL
         Real, Dimension(MAXTS)        :: TotPFSG
         Double Precision              :: RATOUT,HH
      Double Precision, Dimension(MAXRIP,MAXPOLY,MAXTS)  :: ETR
|
         DATA TEXT /'       RIPARIAN ET'/
|     ***********************************************************************  '
|
|
|1-----Set pointers for current grid.
|
      Call SGWF2RIP3PNT(IGRID)
|
|2-----Initialize the rate accumulator (RATOUT)
|
      Zero=0.0
         Ratout=zero
|
|3-----Set cell-by-cell budget save flag (IBD) and clear buffer
|
         IBD=0
         If(IRIPCB < 0 .and. ICBCFL /= 0) IBD=-1
         If(IRIPCB > 0) IBD=ICBCFL
|
|4-----If cell-by-cell flows will be saved as a list, Write header.
|
      IF(IBD.EQ.2) THEN
               CALL UBDSV4(KSTP,KPER,TEXT,NAUX,RIPAUX,IRIPCB,NCOL,NROW,NLAY,   &
               NRIPCL,IOUT,DELT,PERTIM,TOTIM,IBOUND)
      END IF
|
|5-----Initialize BUFF
|
         Do IL=1, NLAY
           Do IR=1, NROW
             Do IC=1, NCOL
                BUFF(IC,IR,IL)=zero
             End do
           End do
         End do
|
|6-----Loop through each riparian cell if NRIPCL > 0.
|
      If(NRIPCL == 0) Return
      Do LC = 1, NRIPCL
|
|7-----Get row, column, layer, and surface elevation.
|
               IL=CLoc(LC,1)
               IR=CLoc(LC,2)
               IC=CLoc(LC,3)
|
|8-----Set HH=HNEW, and initialize RATE to zero
|
               HH=HNEW(IC,IR,IL)
               RATE=zero
|
|9-----Loop through the polygons
|
         Do LP=1,NPoly(LC)
|
```

```
|10---Loop through plant functional subgroups, calculate ETR and
|      aggregate ETR to RATE
|
                      Do NTS=1,MAXTS
                          ETR(LC,LP,NTS)=C1(LC,LP,NTS)*HH+C2(LC,LP,NTS)
                          RATE=RATE+ETR(LC,LP,NTS)
            End do
          End do    |Poly loop
|
|11---Load RATE to budget accumulator and to BUFF and RIPET
|
          RATOUT=RATOUT-RATE
          BUFF(IC,IR,IL)=BUFF(IC,IR,IL)+RATE
              RIPET(LC)=RATE
|
|12---If saving cell-by-cell flows to list, Write flow.
|
          IF(IBD.EQ.2) CALL UBDSVB(IRIPCB,NCOL,NROW,IC,IR,IL,RATE, &
                          AUXVAL,1,0,1,IBOUND,NLAY)
|
        End do    |Cell loop
|
|13---Calculate the total ET for each PFSG.
|
      Do NTS=1,MAXTS
            Sum=0.0
                Do LC=1,NRIPCL
                  Do LP=1,NPOLY(LC)
                      Sum=Sum+ETR(LC,LP,NTS)
                  End do
                End Do
                TotPFSG(NTS)=Sum
          End Do
|
|14---Print locations, polygon number, HSurf, ET rate by plant functional
|      groups, and total ET rate per cell, if requested.
|
      If(IBD < 0)Then
          Write(IOUT,100) TEXT,KPER,KSTP
100       Format(1x,/,1x,A,'    PERIOD',I3,'     STEP',I3)
|
|15---Write out ET rates list file.
|
                  Write(IOUT,120) (RIPNM(II), II=1,MAXTS)
120               format(1x,/,1x,' RIPARIAN ET',/                          &
                          1x,'LAYER  ROW  COLUMN      HEAD       CELL      ',    &
                          'POLYGON    SURFACE                  CELL ET RATE BY ',  &
                          'PLANT FUNCTIONAL GROUP',/                        &
                  1x,'                              ET RATE    NUMBER',   &
                          '     ELEVATION        ',20(2x,A24),/)
            DO LC=1, NRIPCL
                    IL=CLoc(LC,1)
                    IR=CLoc(LC,2)
                    IC=CLoc(LC,3)
                    HH=HNEW(IC,IR,IL)
              DO LP=1,NPOLY(LC)
                If(LP.EQ.1) Then
                Write(IOUT,130) IL,IR,IC,HH,RIPET(LC)
130                   Format(1x,I3,3x,I3,3x,I3,5x,F10.2,1x,F10.5)
                  Write(IOUT,140) LP,HSURF(LC,LP),(ETR(LC,LP,II),II=1,MAXTS)
140             Format(1x,45x,I3,3x,F10.2,9x,20(2x,F10.5,14x))
                Else
                Write(IOUT,140) LP,HSURF(LC,LP),(ETR(LC,LP,II),II=1,MAXTS)
                      End If
              End Do
            End Do
                  Write(IOUT,150) (TotPFSG(II),II=1,MAXTS)
```

```
150        Format(/,1x,32x,'Total ET per PFSG',21x,20(2x,F10.5,14x))
      End If
!
!16---Check to see if cell-by-cell flows are to be saved as three-
!     dimensional array.
!
          If(IBD == 1) Then
                Call UBUDSV(KSTP,KPER,TEXT,IRIPCB,BUFF,NCOL,NROW,NLAY,IOUT)
      End if
!
!17---Check to see if cell location, polygon, head, surface elevation, and
!     plant functional subgroup ET rates are to b saved as a formatted file.
!
      If(IRIPCB1>0) Then
        DO LC=1,NRIPCL
                IL=CLoc(LC,1)
                IR=CLoc(LC,2)
                IC=CLoc(LC,3)
                NP=NPoly(LC)
                HH=HNEW(IC,IR,IL)
          DO LP=1,NP
            Write(IRIPCB1,160) IL,IR,IC,LP,HH,HSurf(LC,LP),     &
                                        (ETR(LC,LP,II),II=1,MAXTS)
160            Format(4I5,2F10.2,25F10.5)
          End Do
        End Do
      End If
!
!18---Move total riparian loss into VBVL for printing in BAS1OT.
!
          VBVL(3,MSUM)=Zero
          VBVL(4,MSUM)=RATOUT
!
!19---Add riparian ET (riparian ET rate times time's step length)
!     to VBVL.
!
          VBVL(2,MSUM)=VBVL(2,MSUM)+RATOUT*DELT
!
!20---Move budget term labels to VBNM for printing by module BAS1OT.
!
      VBNM(MSUM)=TEXT
!
!21---Increment Budget term counter.
!
          MSUM=MSUM+1
!
!22---Return
!
          Return
!
          End Subroutine GWF2RIP3BD
```

## 15.11 HCOFtrm and RHStrm

The functions HCOFtrm and RHStrm are based on a linear interpolation of the ET canopy flux function, $R(h)$, is given by the equation,

$$R(h) = R_k + \frac{(h - h_k)}{h_{k+1} - h_k}(R_{k+1} - R_k) \text{ for } h_k < h \leq h_{k+1},$$

which can be rewritten as

$$R(h) = \frac{R_{k+1} - R_k}{h_{k+1} - h_k}h + \frac{(R_k h_{k+1} - R_{k+1} h_k)}{h_{k+1} - h_k} \text{ for } h_k < h \leq h_{k+1}.$$

The function HCOFtrm is calculated as $\text{HCOFtrm}(h_k, h_{k+1}, R_k, R_{k+1}) = \frac{R_{k+1} - R_k}{h_{k+1} - h_k}$.

The function RHStrm is calculated as $\text{RHStrm}(h_k, h_{k+1}, R_k, R_{k+1}) = \frac{(R_k h_{k+1} - R_{k+1} h_k)}{h_{k+1} - h_k}$.

The variables used in HCOFtrm and RHStrm are listed in table 11.

**Table 11.** List of variables in functions HCOFtrm and RHStrm.

| Variable | Scope | Definition |
|---|---|---|
| HK | Function | First of two consecutive vertices head values |
| HK1 | Function | Second vertex head value |
| RK | Function | First of two consecutive vertices ET flux values |
| RK1 | Function | Second vertex ET flux value |

## 15.12 Listing for HCOFtrm and RHStrm Functions

```
|      ****************************************************************************
       Function HCOFtrm(HK,HK1,RK,RK1)
|      ****************************************************************************
|      This function calculates the HCOF term for the riparian ET
|        HK = first of two consecutive vertices head values
|        HK1 = second vertex head value
|        RK = first of two consecutive vertices ET flux values
|        RK1 = second vertex ET flux value
|      ****************************************************************************
          Double Precision:: HCOFtrm
          Double Precision:: HK,HK1,RK,RK1
|      ****************************************************************************
|
          HCOFtrm=(RK1-RK)/(HK1-HK)
|
       End Function HCOFtrm
|
|      ****************************************************************************
       Function RHStrm(HK,HK1,RK,RK1)
|      ****************************************************************************
|      This function calculates the RHS term for the riparian ET
|        HK = first of two consecutive vertices head values
|        HK1 = second vertex head value
|        RK = first of two consecutive vertices ET flux values
|        RK1 = second vertex ET flux value
|      ****************************************************************************
          Double Precision:: RHStrm
          Double Precision::HK,HK1,RK,RK1
|      ****************************************************************************
|
          RHStrm=(RK*HK1-RK1*HK)/(HK1-HK)
|
       End Function RHStrm
```

## 15.13  Listing for GWF2RIP3DA

```
*********************************************************************
Subroutine GWF2RIP3DA(IGRID)
*********************************************************************
Deallocate RIP-ET memory
*********************************************************************
USE GWFRIPMODULE
*********************************************************************

Deallocate(GWFRIPDAT(IGRID)%MAXRIP)
Deallocate(GWFRIPDAT(IGRID)%MAXPOLY)
Deallocate(GWFRIPDAT(IGRID)%IRIPCB)
Deallocate(GWFRIPDAT(IGRID)%IRIPCB1)
Deallocate(GWFRIPDAT(IGRID)%NRIPCL)
Deallocate(GWFRIPDAT(IGRID)%MAXTS)
Deallocate(GWFRIPDAT(IGRID)%MXSEG)
Deallocate(GWFRIPDAT(IGRID)%RIPNM)
Deallocate(GWFRIPDAT(IGRID)%NuSeg)
Deallocate(GWFRIPDAT(IGRID)%Sxd)
Deallocate(GWFRIPDAT(IGRID)%Ard)
Deallocate(GWFRIPDAT(IGRID)%Rmax)
Deallocate(GWFRIPDAT(IGRID)%Rsxd)
Deallocate(GWFRIPDAT(IGRID)%fdh)
Deallocate(GWFRIPDAT(IGRID)%fdR)
Deallocate(GWFRIPDAT(IGRID)%CLoc)
Deallocate(GWFRIPDAT(IGRID)%NPoly)
Deallocate(GWFRIPDAT(IGRID)%HSurf)
Deallocate(GWFRIPDAT(IGRID)%CovPFSG)
Deallocate(GWFRIPDAT(IGRID)%C1)
Deallocate(GWFRIPDAT(IGRID)%C2)
Deallocate(GWFRIPDAT(IGRID)%RIPET)

Return

End Subroutine GWF2RIP3DA
```

## 15.14  Listing for SGWF2RIP3PNT

```
| ***************************************************************************
  Subroutine SGWF2RIP3PNT(IGRID)
| ***************************************************************************
  Set pointers to RIP-ET data for grid.
| ***************************************************************************
  USE GWFRIPMODULE
| ***************************************************************************
  MAXRIP   => GWFRIPDAT(IGRID)%MAXRIP
  MAXPOLY  => GWFRIPDAT(IGRID)%MAXPOLY
  IRIPCB   => GWFRIPDAT(IGRID)%IRIPCB
  IRIPCB1  => GWFRIPDAT(IGRID)%IRIPCB1
  NRIPCL   => GWFRIPDAT(IGRID)%NRIPCL
  MAXTS    => GWFRIPDAT(IGRID)%MAXTS
  MXSEG    => GWFRIPDAT(IGRID)%MXSEG
  RIPNM    => GWFRIPDAT(IGRID)%RIPNM
  NuSeg    => GWFRIPDAT(IGRID)%NuSeg
  Sxd      => GWFRIPDAT(IGRID)%Sxd
  Ard      => GWFRIPDAT(IGRID)%Ard
  Rmax     => GWFRIPDAT(IGRID)%Rmax
  Rsxd     => GWFRIPDAT(IGRID)%Rsxd
  fdh      => GWFRIPDAT(IGRID)%fdh
  fdR      => GWFRIPDAT(IGRID)%fdR
  CLoc     => GWFRIPDAT(IGRID)%CLoc
  NPoly    => GWFRIPDAT(IGRID)%NPoly
  HSurf    => GWFRIPDAT(IGRID)%HSurf
  CovPFSG  => GWFRIPDAT(IGRID)%CovPFSG
  RIPET    => GWFRIPDAT(IGRID)%RIPET
  C1       => GWFRIPDAT(IGRID)%C1
  C2       => GWFRIPDAT(IGRID)%C2
|
  Return
|
  End Subroutine SGWF2RIP3PNT
```

## 15.15 Listing for SGWF2RIP3PSV

```
*******************************************************************************
Subroutine SGWF2RIP3PSV(IGRID)
*******************************************************************************
Save pointers to RIP data for grid.
*******************************************************************************
USE GWFRIPMODULE
*******************************************************************************
GWFRIPDAT(IGRID)%MAXRIP   => MAXRIP
GWFRIPDAT(IGRID)%MAXPOLY  => MAXPOLY
GWFRIPDAT(IGRID)%IRIPCB   => IRIPCB
GWFRIPDAT(IGRID)%IRIPCB1  => IRIPCB1
GWFRIPDAT(IGRID)%NRIPCL   => NRIPCL
GWFRIPDAT(IGRID)%MAXTS    => MAXTS
GWFRIPDAT(IGRID)%MXSEG    => MXSEG
GWFRIPDAT(IGRID)%RIPNM    => RIPNM
GWFRIPDAT(IGRID)%NuSeg    => NuSeg
GWFRIPDAT(IGRID)%Sxd      => Sxd
GWFRIPDAT(IGRID)%Ard      => Ard
GWFRIPDAT(IGRID)%Rmax     => Rmax
GWFRIPDAT(IGRID)%Rsxd     => Rsxd
GWFRIPDAT(IGRID)%fdh      => fdh
GWFRIPDAT(IGRID)%fdR      => fdR
GWFRIPDAT(IGRID)%CLoc     => CLoc
GWFRIPDAT(IGRID)%NPoly    => NPoly
GWFRIPDAT(IGRID)%HSurf    => HSurf
GWFRIPDAT(IGRID)%CovPFSG  => CovPFSG
GWFRIPDAT(IGRID)%RIPET    => RIPET
GWFRIPDAT(IGRID)%C1       => C1
GWFRIPDAT(IGRID)%C2       => C2

Return

End Subroutine SGWF2RIP3PSV
```

# 16.0 References Cited

Ajami, Hoori, Maddock,T., III, Meixner, T., Hogan, J.F., Guertin, D.P., 2011, RIPGIS-NET: A GIS Tool for Riparian Groundwater Evapotranspiration in MODFLOW: Ground Water, Methods/Note, 5 p., doi: 10.1111/j.1745-6584.2011.00809.x.

Baird, K.J., and Maddock, T., III, 2005, Simulating riparian evapotranspiration: a new methodology and application for groundwater models: Journal of Hydrology, v. 312, p. 176-190.

Banta, E.R., 2000, MODFLOW-2000, the U.S. Geological Survey modular ground-water model—Documentation of packages for simulating evapotranspiration with a segmented function (ETS1) and drains with return flow (DRT1): U.S. Geological Survey Open-File Report 00-466.

Barth, G., Barroll, P., Hathaway, D.L., Maddock III, T., Shafike, N., King, J.P., Shomaker, J., and Liu, B., 2008, Building a new groundwater flow model for the Rincon and Mesilla Bolsons: Modflow and More - Ground Water and Public Policy, Golden, Colorado, May 18-21, 2008, p. 252-256.

Berger, D.L., Johnson, M.J., and Tumbusch, M.L., 2001, Estimates of evapotranspiration from the Ruby Lake National Wildlife Refuge Area, Ruby Valley, northeastern Nevada, May 1999–October 2000: U.S. Geological Survey Water-Resources Investigations Report 01-4234.

Busch, D.E., Ingraham, N.L., and Smith, S.D., 1992, Water uptake in woody riparian phreatophytes of the southwestern United States: A stable isotope study: Ecological Applications, v. 2, p. 450-459.

Glennon, R.J., and Maddock, T., III, 1994, In search of subflow: Arizona's futile effort to separate groundwater from surface water: 36 Arizona Law Review 567.

Goodrich, D.C., Scott, R., Qi, J., Goff, B., Unkrich, C.L., Moran, M.S., Williams, D.G., Schaeffer, S.M., Snyder, K.A., MacNish, R.D., Maddock, T., Pool, D., Chehbouni, A., Cooper, D.I., Eichinger, W.E., Shuttleworth, W.J., Kerr, Y., Marsett, R., and Ni, W., 2000, Seasonal estimates of riparian evapotranspiration using remote and in situ measurements: Agricultural and Forest Meteorology, v. 105, p. 281-309.

Granier, A., Biron, P., and Lemoine, D., 2000, Water balance, transpiration and canopy conductance in two beech stands: Agricultural and Forest Meteorology, v. 100, p. 291-308.

Grimm, N.B., Chacon, A., Dahm, C.N., Hostetler, S.W., Lind, O.T., Starkweather, P.L., and Wurtsbaugh, W.W., 1997, Sensitivity of aquatic ecosystems to climatic and anthropogenic changes: the Basin and Range, American Southwest and Mexico: Hydrological Processes, v. 11, p. 1023-1041.

Hadley, G., 1964, Nonlinear and Dynamic Programming: Addison-Wesley Publishing Co. Inc., Reading Massachusetts.

Harbaugh, A.W., 2005, MODFLOW-2005, The U.S. Geological Survey Modular Ground-Water Model—the Ground-Water Flow Process: U.S. Geological Survey Techniques and Methods 6-A16, various pagination. (Also available at http://pubs.usgs.gov/tm/2005/tm6A16/PDF.htm.)

Harbaugh, A.W., and McDonald, M.G., 1996a, User's documentation for MODFLOW-96: An update to the U.S. Geological Survey modular finite-difference ground-water flow model: U.S. Geological Survey Open-File Report 96–485.

Harbaugh, A.W., and McDonald, M.G., 1996b, Programmer's documentation for MODFLOW-96, an update to the U.S. Geological Survey modular finite difference ground-water flow model: U.S. Geological Survey Open-File Report 96–486.

Kimura, R., Bai, L., Fan, J., Takayama, N., and Hinokididani, O., 2007, Evapotranspiration estimates over the river basin of the Loess Plateau of China based on remote sensing: Journal of Arid Environments, v. 68, issue 1, p. 53-65.

Lavorel, S., McIntyre, S., Landsberg, J.J., and Forbes, T.D.A., 1997, Plant functional classifications: from general groups to specific groups based on response to disturbance: Tree, v. 12, p. 474-478.

Leishman, M.R., and Westoby, M., 1992, Classifying plants into groups on the basis of associations of individual traits – evidence from Australian semi-arid woodlands: Journal of Ecology, v. 80, p. 417-424.

Maddock, T., III, and Baird, K.J., 2003, A riparian evapotranspiration package for MODFLOW-96 and MODFLOW-2000: HWR No. 02-03, Department of Hydrology and Water Resources, University of Arizona Research Laboratory for Riparian Studies, University of Arizona, Tucson, Arizona.

McDonald, M.G., and Harbaugh, A.W., 1988, A Modular Three-Dimensional Finite-Difference Ground-Water Flow Model: U.S. Geological Survey Techniques of Water Resources Investigations 6-A1.

Meinzer, F.C., Andrade, J.L., Goldstein, G., Holbrook, N.M., Cavelier, J., and Jackson, P., 1997, Control of transpiration from the upper canopy of a tropical forest: the role of stomatal layer and hydraulic architecture components: Plant, Cell and Environment, v. 20, p. 1242-1252.

Nagler, P.L., Scott, R.L., Westenburg, C., Cleverly, J.R., Glen, E.P., and Huete, A.R., 2005, Evapotranspiration on western U.S. rivers estimated using the Enhanced Vegetation Index from MODIS and data from eddy covariance and Bowen ratio flux towers: Remote Sensing of Environment, v. 97, p. 337-351.

Oren, R., Phillips, N., Ewers, B.E., Pataki, D.E., and Megonigal, J.P., 1999, Sap-flux-scaled transpiration responses to light, vapor pressure deficit, and leaf area reduction in a flooded Taxodium distichum forest: Tree Physiology, v. 19, p. 337-347.

Poff, N.L., Allan, J.D., Bain, M.B., Karr, J.R., Prestegaard, K.L., Richter, B.D., and Stromberg, J.C., 1997, The natural flow regime: A paradigm for river conservation and restoration: BioScience, v. 47, p. 769-784.

Rana, G., and Katerji, N., 2000, Measurement and estimation of actual evapotranspiration in the field under Mediterranean climate: a review: European Journal Agronomy, v. 13, issues 2-3, p. 125-153.

Schmid, Wolfgang, and Hanson, R.T., 2009, The Farm Process Version 2 (FMP2) for MODFLOW-2005 - Modifications and Upgrades to FMP1: U.S. Geological Survey Techniques and Methods 6-A32, 102 p. (Also available at http://pubs. usgs.gov/tm/tm6a32/.)

Schmid, Wolfgang, Hanson, R.T., Maddock III, T.M., and Leake, S.A., 2006, User's guide for the Farm Package (FMP1) for the U.S. Geological Survey's modular three-dimensional finite-difference groundwater flow model, MODFLOW-2000: U.S. Geological Survey Techniques and Methods 6-A17, 127 p.

Schorr, S.W., 2005, Hydrologic Assessment and Simulations of Groundwater Conditions in Arivaca Basin, Pima County, Arizona: University of Arizona, M.S. thesis.

Scott, R., Edwards, E., Shuttleworth, J., Huxman, T., Watts, C., and Goodrich, D., 2004, Interannual and seasonal variation in fluxes of water and carbon dioxide from a riparian woodland ecosystem: Agricultural and Forest Meteorology, v. 122, p. 65-84.

Shafroth, P.B., Stromberg, J.C., and Patten, D.T., 2000, Woody riparian vegetation response to different alluvial water table regimes: Western North American Naturalist, v. 60, no. 1, p. 66-76.

Snyder, K.A., and Williams, D.G., 2000, Water sources used by riparian trees varies among stream types on the San Pedro River, Arizona: Agricultural and Forest Meteorology, v. 105, p. 227-240.

Steinwand, A.L., Harrington, R.F., and Groeneveld, D.P., 2001, Transpiration coefficients for three Great Basin shrubs: Journal of Arid Environments, v. 49, p. 555–567.

Stromberg, J.C., 1993, Fremont cottonwood-Goodding willow riparian forests: A review of their ecology, threats, and recovery potential: Journal of the Arizona-Nevada Academy of Science, v. 26, no. 3, p. 97-110.

Stromberg, J.C., 2001, Restoration of riparian vegetation in the southwestern United States: importance of flow regimes and fluvial dynamism: Journal of Arid Environments, v. 49, p. 17-34.

Stromberg, J.C, Tiller, R., and Richter, B., 1996, Effects of groundwater decline on riparian vegetation of semiarid regions: the San Pedro, Arizona: Ecological Applications, v. 6, no. 1, p. 113-131.

Vreugdenhil, S.J., Kramer, K., and Pelsma, T., 2006, Effects of flooding duration, frequency and depth on the presence of saplings of six woody species in north-west Europe: Forest Ecology and Management, v. 236, p. 47-55.

Williams, D.G., Brunel, J.P., Schaeffer, S.M., and Snyder, K.A., 1998, Biotic controls over the functioning of desert riparian ecosystems, in Wood, E.F., Chehbouni, A.G., Goodrich, D.C., Seo, D.J., and Zimmerman, J.R., eds., Proceedings from the Special Symposium on Hydrology: American Meteorological Society, p. 43-48.

# Appendix A.  Plant Functional Subgroup Curves

Listed here are the heads, transpiration fluxes, fdh, and fdR for the plant functional groups developed for RIP-ET for the South Fork of the Kern River, California. The values are the result of fieldwork, literature review, and researcher input (Baird and Maddock, 2005). The head is defined as zero at the land surface, positive above land surface, and negative below land surface. The units of head are centimeters and the units of transpiration flux are centimeters per day for all curves. In the RIP-ET package, vertex 1 and segment 1 of the flux curve must start at the PFSG's extinction depth. Therefore, fdh and fdR array indices must start at that point as well.

**Table A1.**   Head, tranpiration fluxes for obligate wetland PFG.

[cm, centimeter; cm/d, centimeter per day]

| Vertex | Head (cm) | Flux (cm/d) |
|--------|-----------|-------------|
| 1 | −75 | 0 |
| 2 | −33 | 0.20 |
| 3 | −15 | 0.32 |
| 4 | −10 | 0.36 |
| 5 | −7 | 0.35 |
| 6 | −4 | 0.32 |
| 7 | 0 | 0.23 |
| 8 | 20 | 0 |

**Table A2.**   fdh and fdR for obligate wetland PFG.

| Segment | fdh | fdR |
|---------|-----|-----|
| 1 | 0.4421 | 0.5556 |
| 2 | 0.1895 | 0.3333 |
| 3 | 0.0526 | 0.1111 |
| 4 | 0.0316 | −0.0278 |
| 5 | 0.0316 | −0.0833 |
| 6 | 0.0421 | −0.2500 |
| 7 | 0.2105 | −0.6389 |

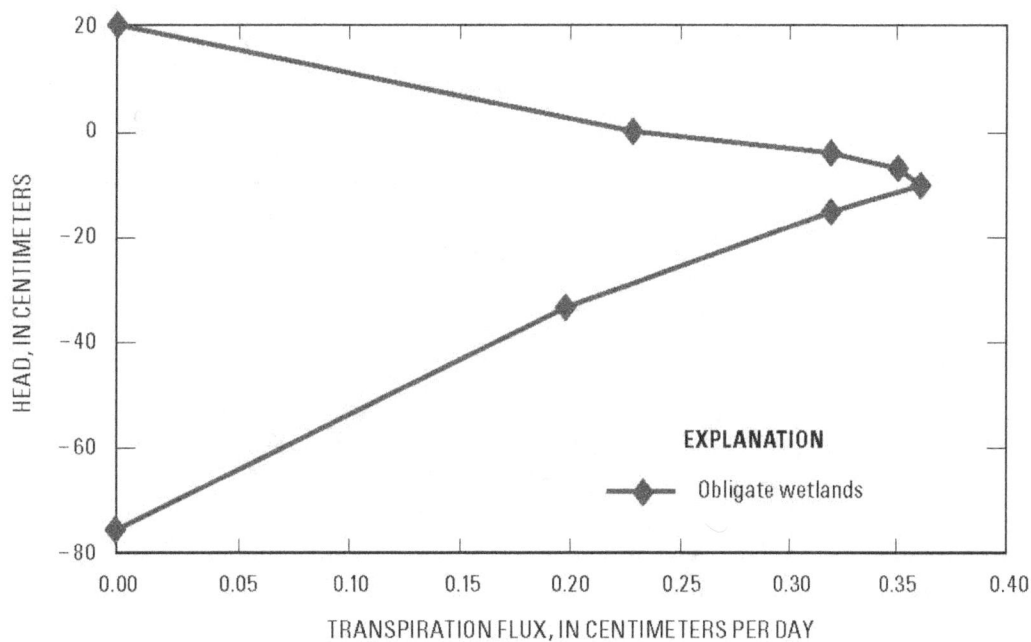

**Figure A1.**   Transpiration fluxes for obligate wetland PFG.

**Table A3.**   Head, transpiration fluxes for shallow-rooted riparian PFG.

[cm, centimeter; cm/d, centimeter per day]

| Vertex | Head (cm) | Flux (cm/d) |
|--------|-----------|-------------|
| 1 | −160 | 0 |
| 2 | −100 | 0.10 |
| 3 | −60 | 0.31 |
| 4 | −50 | 0.33 |
| 5 | −40 | 0.31 |
| 6 | −32 | 0.17 |
| 7 | 0 | 0 |

**Table A4.**   fdh and fdR for shallow-rooted riparian PFG.

| Segment | fdh | fdR |
|---------|------|------|
| 1 | 0.3750 | 0.3030 |
| 2 | 0.2500 | 0.6364 |
| 3 | 0.0625 | 0.0606 |
| 4 | 0.0625 | −0.0606 |
| 5 | 0.0500 | −0.4242 |
| 6 | 0.2000 | −0.5152 |

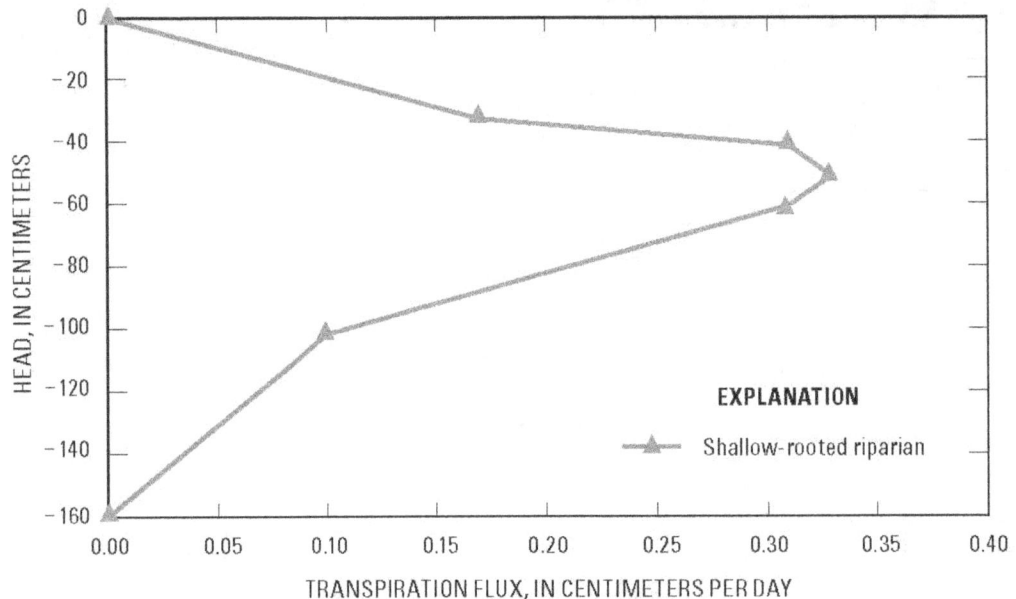

**Figure A2.**   Transpiration fluxes for shallow-rooted riparian PFG.

**Table A5.** Head, transpiration fluxes for average deep-rooted riparian PFG.

[cm, centimeter; cm/d, centimeter per day]

| Vertex | Head (cm) | Flux (cm/d) |
|--------|-----------|-------------|
| 1 | −500 | 0.00 |
| 2 | −400 | 0.12 |
| 3 | −300 | 0.21 |
| 4 | −200 | 0.29 |
| 5 | −150 | 0.31 |
| 6 | −100 | 0.28 |
| 7 | −75 | 0.17 |
| 8 | 0 | 0.00 |

**Table A6.** fdh and fdR for average deep-rooted riparian PFG.

| Segment | fdh | fdR |
|---------|--------|---------|
| 1 | 0.2000 | 0.3871 |
| 2 | 0.2000 | 0.2903 |
| 3 | 0.2000 | 0.2581 |
| 4 | 0.1000 | 0.0645 |
| 5 | 0.1000 | −0.0968 |
| 6 | 0.0500 | −0.3548 |
| 7 | 0.1500 | −0.5484 |

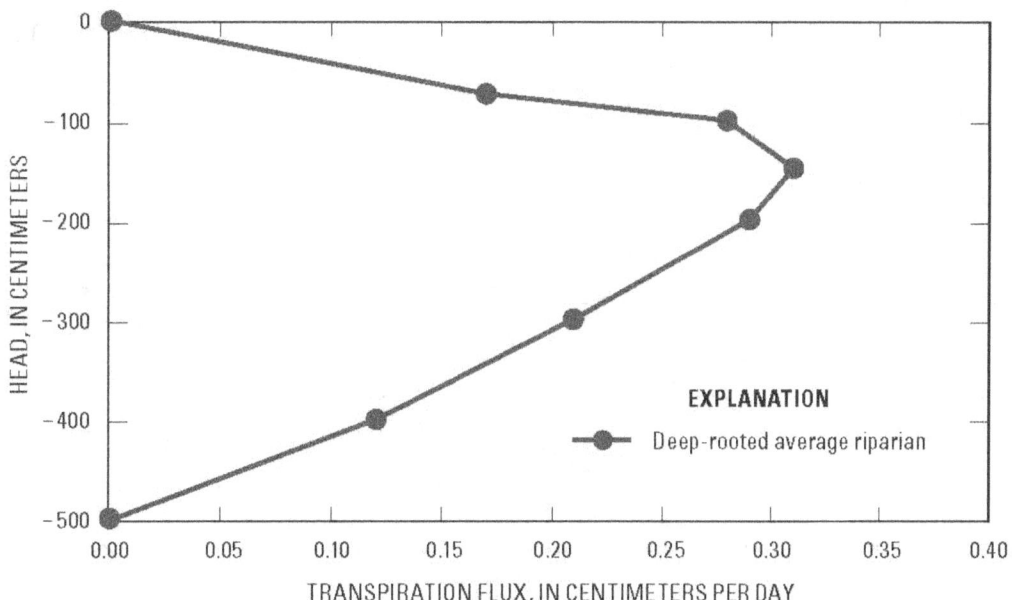

**Figure A3.** Transpiration fluxes for average deep-rooted riparian PFG.

**Table A7.**  Head, transpiration fluxes for transitional riparian PFG.

[cm, centimeter; cm/d, centimeter per day]

| Vertex | Head (cm) | Flux (cm/d) |
|--------|-----------|-------------|
| 1 | −600 | 0 |
| 2 | −500 | 0.11 |
| 3 | −400 | 0.20 |
| 4 | −340 | 0.27 |
| 5 | −300 | 0.28 |
| 6 | −260 | 0.27 |
| 7 | −175 | 0.16 |
| 8 | −50 | 0.00 |

**Table A8.**  fdh and fdR for transitional riparian PFG.

| Segment | fdh | fdR |
|---------|-----|-----|
| 1 | 0.1818 | 0.3929 |
| 2 | 0.1818 | 0.3371 |
| 3 | 0.1091 | 0.2343 |
| 4 | 0.0727 | 0.0357 |
| 5 | 0.0727 | −0.0357 |
| 6 | 0.1545 | −0.3811 |
| 7 | 0.2273 | −0.5832 |

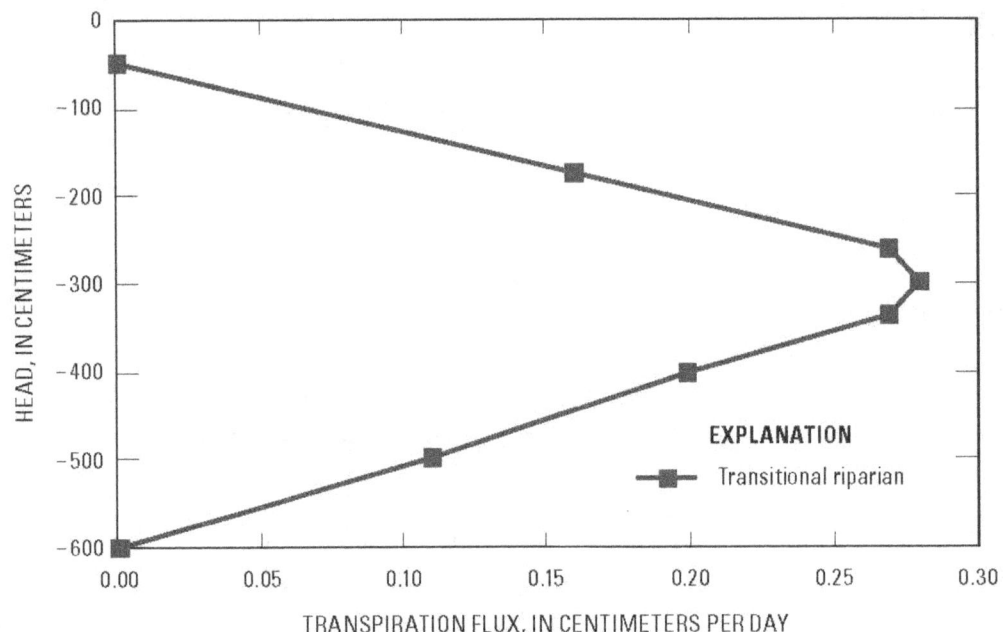

**Figure A4.**  Transpiration fluxes for transitional riparian PFG.

**Table A9.**   Head, transpiration flux for deep-rooted riparian size classes.

[cm, centimeter; cm/d, centimeter per day]

| Head (cm) | Flux (cm/d) | | | |
|---|---|---|---|---|
| | Small | Medium | Large | X-Large |
| -500 | | 0 | 0 | 0 |
| -400 | 0 | 0.0838 | 0.1032 | 0.1100 |
| -300 | 0.0501 | 0.1650 | 0.2064 | 0.2199 |
| -200 | 0.1001 | 0.2450 | 0.2759 | 0.2940 |
| -150 | 0.1602 | 0.2615 | 0.2944 | 0.3137 |
| 100 | 0.2003 | 0.2357 | 0.2653 | 0.2827 |
| 75 | 0.1742 | 0.1506 | 0.1400 | 0.1492 |
| 0 | 0 | 0 | 0 | 0 |

**Table A10.**   `fdh` and `fdR` for deep-rooted riparian size classes.

| Segment number | Small | | Medium | | Large | | X-large | |
|---|---|---|---|---|---|---|---|---|
| | `fdh` | `fdR` | `fdh` | `fdR` | `fdh` | `fdR` | `fdh` | `fdR` |
| 1 | 0.2500 | 0.2501 | 0.2000 | 0.3218 | 0.2000 | 0.3505 | 0.2000 | 0.3503 |
| 2 | 0.2500 | 0.2496 | 0.2000 | 0.3103 | 0.2000 | 0.3505 | 0.2000 | 0.3503 |
| 3 | 0.1250 | 0.3001 | 0.2000 | 0.3066 | 0.2000 | 0.2362 | 0.2000 | 0.2382 |
| 4 | 0.1250 | 0.2002 | 0.1000 | 0.0613 | 0.1000 | 0.0628 | 0.1000 | 0.0612 |
| 5 | 0.0625 | −0.1303 | 0.1000 | −0.0958 | 0.1000 | −0.0988 | 0.1000 | −0.0986 |
| 6 | 0.1875 | −0.8697 | 0.0500 | −0.3257 | 0.0500 | −0.4256 | 0.0500 | −0.4252 |
| 7 | | | 0.1500 | −0.5785 | 0.1500 | −0.4756 | 0.1500 | −0.4762 |

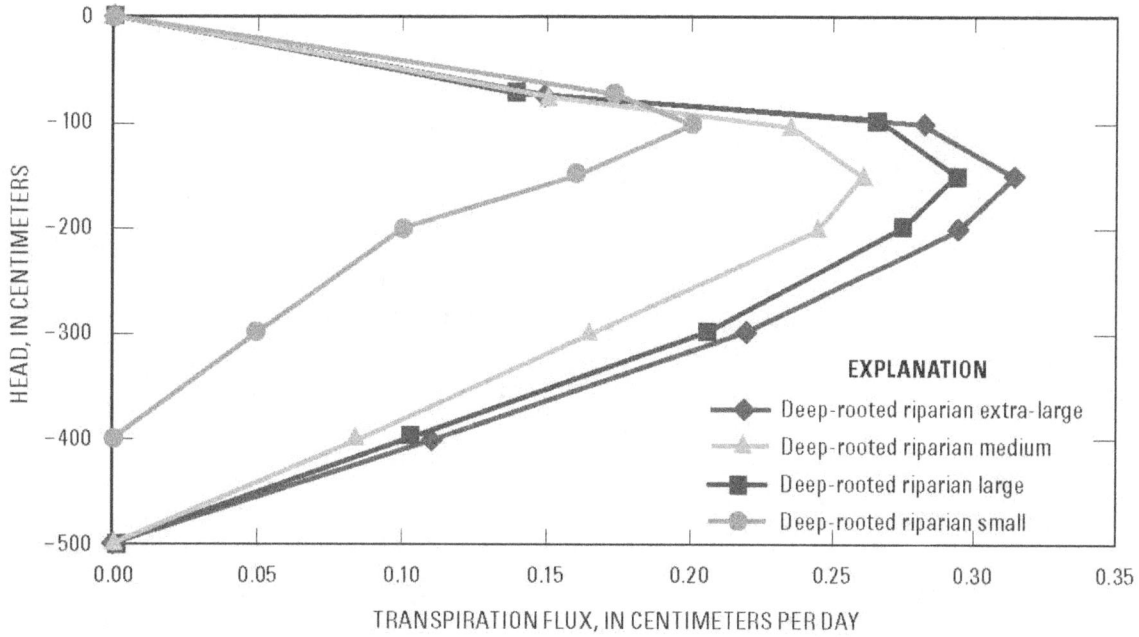

**Figure A5.**   Transpiration fluxes for small, medium, large and X-large deep-rooted riparian PFG.